BASIC
ENGINEERING MECHANICS
Explained

The foundation of engineering knowledge

Volume 2
Motion and Energy

Gregory Pastoll, PhD (Higher Ed) BSc Mech Eng

Copyright and Origination

The contents of this work are entirely the work of this author, who retains the copyright in them.

Besides a few historical images that are in the public domain, all the illustrations are the work of the author.

This work was first published by Gregory Pastoll, in 2019

Cover design by Ross MacLennan

Cover artwork by Gregory Pastoll

ISBN   978-0-6484665-2-9 (hardcover)

ISBN   978-0-6484665-3-6 (paperback)

ISBN   978-0-6484665-7-4 (e-book)

Preface

I have written this book because, as someone who has taught engineering mechanics enthusiastically for over 14 years, I believe that mechanics is interesting, easy and relevant to many aspects of our daily life.

None of the books on mechanics that I have seen give that impression, so this book is an attempt to redress that issue. I have tried here to present the material in a way that makes the content self-evidently useful and appealing.

As a young lecturer in basic mechanical engineering subjects at the former Cape Technikon, in Cape Town, South Africa, I experimented with teaching methods, and began to develop an understanding of what was needed to give students a practical 'handle' on what they were learning.

While teaching there, I was greatly inspired by the successes achieved by John Cowan, a visiting lecturer, in making university learning in the field of engineering truly 'hands-on'. At the time, Cowan was a lecturer in civil engineering at Heriot-Watt University in Scotland. He later went on to become Professor of Learning Development at The Open University of the UK. After attending his talk in Cape Town, I read some of his publications, and later visited his unit. I could see in the way he taught, that it was possible for students to learn about principles and techniques by taking responsibility for their own learning. Cowan's dedication to this ideal inspired me to explore what was really important in my teaching, and to keep on developing better ways of teaching.

Subsequent to my first 5-year spell of lecturing basic engineering subjects, for a further 14 years I was a consultant in teaching methods at the University of Cape Town. There I was directly exposed to the teaching methods in use in all the different Faculties of the University, and learnt a great deal from trying to work with my colleagues towards ways of improving their own teaching.

This knowledge I summarised in a book entitled 'Motivating People to Learn', and I put what I learnt there to good use when I again went into teaching engineering mechanics, for a second spell of 9 years.

The present book is intended to give the the reader the opportunity to:

- Remain interested in and motivated by the subject,
- See how these principles can be applied to a wide range of practical situations,
- Learn to reason out solutions to new problems from first principles, and
- Develop a perspective on the limitations of determining answers by calculations.

The topics covered in the book do not necessarily follow any particular prescribed curriculum. The book contains the essential topics in mechanics found in most curricula, simply because those topics undeniably form the basis of the science of mechanics.

I have added what I perceive to be context-enriching material, about certain historical aspects of the science of mechanics. Rather than keep the content to a minimum, I have opted to include everything I can think of that might broaden a reader's understanding. This has made it necessary to spread the book over three volumes.

Acknowledgements

Professor John Cowan for his many hours spent in patient perusal of my developing chapters and giving me excellent feedback.

Mark Kilfoil for occasional checking and verifying of approaches and calculations.

My mother, Lois, for believing in me since I was born. My late father, Gerald, for showing his children by example always to question everything and to think things through for themselves.

Nick Hampton for telling me that I make it look simpler than it really is, which amused me greatly, as I believe that such a view comes from other authors having created the impression that mechanics is more complicated than it really is.

Those of my ex-colleagues who, over the years, bemusedly, but loyally and voluntarily, assisted me in the often wearisome task of adjudicating some of the design-and-build projects that I assigned to my students. There are too many who helped in some way for me to recall everyone. However, particular support came from Joseph Basakayi, Ian Noble-Jack, John Byett, Margot Lynn, Dave Evans, Eric Obeng and Mark Ludick.

The late Hugh Williams, a brilliant technician who built many unique pieces of laboratory apparatus for me in my first spell of teaching. A most respected, competent and inspiring gentleman, whose input to my laboratory classes was invaluable.

The support and understanding shown me by my wife Lindsay during the long process of getting the book written and illustrated. Without her, I could not have got this far.

This book is dedicated to the thousands of students who have passed through my classes. Thank you for being there so I could try to work out how better to guide your learning. I hope you had fun, and that you retain good memories of your engagement with mechanics. I do.

Gregory Pastoll

Foreword

by John Cowan, Emeritus Professor of Learning Development, UK Open University

This set of three volumes is the work of an enthusiast. Greg has long approached the subject, the teaching and the facilitation of students' learning of applied mechanics, with an inherent enthusiasm which he generously and infectively shares with his readers. With a keen awareness that learners in this subject area must go beyond *understanding* the concepts and theory to *applying* them, he follows his initial instruction by offering a generous provision of examples. With an experienced teacher's wisdom, he appreciates that such a series could prove wearisome or off-putting for both over-confident or uncertain learners. He thus presents many of his almost standard examples wrapped up for variety in motivating and brief scenario settings.

Motivation and interest have been high on his list of priorities as a writer; so, too, has been the fundamental task of explaining principles, concepts and methods. He covers this comprehensively, competently and in a logical and well-explained sequence. This series has been constructed with meticulous attention paid to the needs of the learners who are the readers. Reader-friendly language is used throughout. New terms are introduced timeously and clearly, but never in the tedious and formal detail which can bedevil many texts on this subject.

Occasionally Greg indulges in the familiar eccentricity of the born teacher by wandering off into a digression in which he tells an interesting "wee story". He airs relevant comments and recounts interesting historical accounts or anecdotes, whose mastery may not be necessary in grasping the fundamental principles and approaches, but whose value in retaining the interest of the reader while enhancing their relevant knowledge is noteworthy.

From time to time, his writing includes wry comments which brought a smile to the lips of this reader. This contributes subtly to his writer's efforts to establish a relationship with his readers in which he and they engage together in the learning and teaching of applied mechanics. The text is also enriched by the inclusion of many attractive and relevant sketches from Greg's able hand, through which he contributes further to the personal nature of the teacher/student relationship housed in these pages.

The series has been conceived on the basis of extensive and reflectively reviewed teaching in this area coupled with wide reading of the approaches followed by others. For Greg has been enthusiastically seeking to address the assorted needs of learners of varied ability, experience and motivation.

The result is comprehensive and developmental coverage of the principles and practice of applied mechanics. In my judgement, the end result is almost self-sufficient for any learner who wishes to master this subject area by following this text. It lends itself to use in the context which has been called the "flipped classroom".

Here study time begins with time spent assimilating a provided teaching input on text or video. This is then followed by class time in which the teacher, supported by peers, addresses difficulties and questions, and then facilitates the addressing of tasks designed to deepen the learning and consolidate the understanding. This text would be a wonderful source for self-managed learning time, and many of its activities ideal for the follow-up time.

I am somewhat sad that, with my advancing years, the opportunity to build in this way on Greg's work is not open to me; but it is open for today's teachers and today's learners, whom I collegially urge to take advantage of it.

John Cowan

Contents outline for the series

Volume 1: Principles and static forces

1 What Mechanics is about, and why we study it

2 Concepts, quantities, principles and laws

3 Working with numbers in engineering

4 Forces: components, resultants and equilibrium of a particle

5 Force moments, torque, equilibrium of rigid bodies, free-body diagrams

6 Centres of gravity and centroids

7 Forces in structures: frames and trusses

8 Friction between dry flat surfaces

9 Buoyancy

Volume 2: Motion and energy

10 Linear motion with uniform acceleration

11 Motion influenced by gravity: vertical and projectile motion

12 Rotary motion

13 Work, energy and power

14 Simple lifting machines

15 Inertia in linear accelerating systems

16 Linear momentum and impulse

17 Relative velocity

Volume 3: Rotation and inertia

18 Centrifugal and centripetal force

19 Rotational Inertia

20 Rotational and linear inertia in accelerating systems

21 Kinetic energy of rotation and angular momentum

22 Simple harmonic motion

23 Vehicle dynamics

24 Additional exercises, test questions and challenges

Contents: Volume 2

Volume 2: Motion and energy

10 Linear motion with uniform acceleration 1

11 Motion influenced by gravity: vertical and projectile motion 21

12 Rotary motion 45

13 Work, energy and power 67

14 Simple lifting machines 103

15 Inertia in linear accelerating systems 139

16 Linear momentum and impulse 151

17 Relative velocity 193

    Index for Volume 2 219

    About the Author 223

Chapter 10

Linear motion with uniform acceleration

Definitions of the quantities associated with motion

Distinguishing between positive and negative directions

The elementary case of motion with constant speed

Acceleration and deceleration on a velocity-time graph

Interpreting velocity-time graphs

Derivation and application of the three basic equations governing motion with uniform acceleration

Making use of a displacement-time graph (with a practical experiment)

Definitions of the quantities associated with motion

Motion, and the measurement of motion, affects everyone's lives. We walk, run, use bicycles, cars, aircraft and canoes to get somewhere different to where we started from.

We throw balls and javelins, and are interested in the record times attained by athletes, and the speed records attained by vehicles.

The quest for speed: impression of an early Buick racing car

We plan journeys by cars, ships, trains, aircraft and spacecraft. We monitor the velocity of fluids in pipes and canals, and design conveyor systems to deliver products at optimum rates. Knowing about motion is vital for the engineer.

In order to analyse motion, we have to set up ground-rule definitions of the essential quantities involved, namely: distance, displacement, speed, velocity and acceleration.

1

'Distance' is the number of units of length between two given points.

We use the word 'distance' to indicate the extent of the physical separation between two objects. For example, the distance between two mountain peaks or two cities, or two trees.

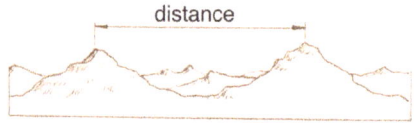

If we are referring to the dimensions of a single object, we talk about its length, width, thickness, depth, radius, diameter or circumference.

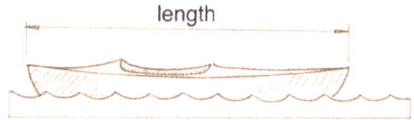

Dimensions of objects are measured in the same units as those for distance.

All dimensions are scalar quantities: A metre rule is a metre long, no matter in which direction it is pointed.

A 'distance' is a scalar quantity, implying that it is not necessary to specify a *direction* in order to define a distance. A distance, such as that between two towns, can be described either in terms of the number of kilometres you have to travel to get there by a winding road, or in terms of the number of kilometres measured in a straight line.

'Displacement' is a measure of the position of an object in relation to a given reference point.

To summarise the difference between distance and displacement, When you ask 'How far?' the answer is a distance, and when you ask 'Where is?' the answer is a displacement.

Suppose you start walking from point A on a straight road, and proceed due east for 5 km, then turn around and walk back along the road, due west for 3 km, stopping at point B. You have covered a total distance of 8 km, and your displacement from point A is now 2 km to the east.

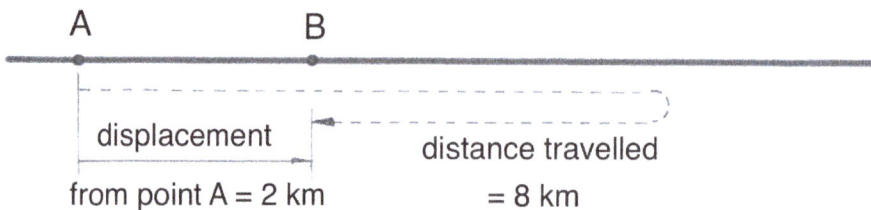

Another example: you could walk right around an athletics track, and when you returned to the starting point, you would have covered a distance of 400 m, but your displacement from your starting point would be zero. It should be clear that distance covered is not the same as displacement. Since a displacement has to be described in terms of both magnitude and direction, displacement is a vector quantity.

Distance travelled = 400 m, but

displacement from starting point = 0 m.

start and
finish

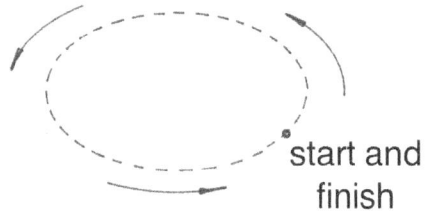

'Speed' (a scalar) is the rate at which an object covers distance.

If you take a trip of 90 km in a car, and accomplish this in one hour, then your average speed on the trip is 90 km/h. Speed is defined as (distance covered ÷ time taken). Since distance is a scalar quantity, speed is also a scalar quantity. It is not dependent on direction. A speed can be experienced in any direction, and it would still be the same speed. You could ride your bicycle at 40 km/h, whether you rode in a circle, or due North, or to town and back, you would still be experiencing a speed of 40 km/h. You experience speed as the rate at which you are moving relative to your immediate surroundings. A child riding a roundabout can be whizzing along at a peripheral speed of, say, 4 m/s, which would be 4 m/s, irrespective of the instantaneous direction in which she is moving.

'Velocity' (a vector) is the instantaneous rate of change of displacement.

Sometimes people use the terms 'speed' and 'velocity' interchangeably, but there is a distinct difference between them. Velocity is the same as speed in one sense, but not in another. Whereas speed is the rate of change of *distance*, velocity is the rate of change of *displacement*. Since displacement is a vector quantity, it follows that velocity is also a vector quantity. The *speed* of an object is identical with the *magnitude* of its velocity. However, to specify a velocity, it is also necessary to specify the *direction* in which this velocity is occurring.

Suppose you are standing next to a straight road, and right in front of you, two cars pass each other in opposite directions, each moving at a speed of 60 km/h. While their speeds are the same, their velocities are different. Relative to the observer at the point on the road where they crossed, one of them has a velocity of 60 km/h to the left, and the other has a velocity of 60 km/h to the right. Defining these respective directions is an essential part of defining a velocity.

'Acceleration' is the rate of change of speed *or* velocity.

Applying this definition to changes in speed: suppose you are a sprinter, running in a race over 100 m. In order to move from being at rest, where your speed is zero, to the point where you reach your top speed, you need to accelerate. The greater your acceleration, the sooner you reach top speed. The same applies to anything that moves, including cars, hot rods and aircraft on take-off.

In the case of changes to an object's velocity, acceleration can refer to the rate of change in *either* the magnitude *or* the direction of the velocity. As an example of the latter is found when a small mass-piece is being swung around in a horizontal circle at the end of a string.

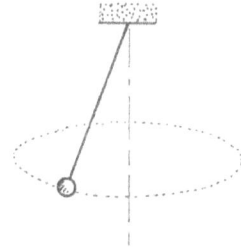

The *speed* of the mass-piece might remain constant, so the *magnitude* of its velocity does not change. However, its *direction* of movement is continually changing, which means the velocity vector keeps changing. This means that the mass-piece is experiencing a form of acceleration.

Although we mention this type of acceleration, in the present chapter we will *not* be dealing with it. Rotational motion will be dealt with in later chapters.

Distinguishing between positive and negative directions

In mathematics and physics, in any situation in which positive signs and negative signs are used, these signs simply designate 'one way' and 'the opposite way'. The convention of positive and negative signs, as applied to a number line, will be used with all vector quantities mentioned in this chapter.

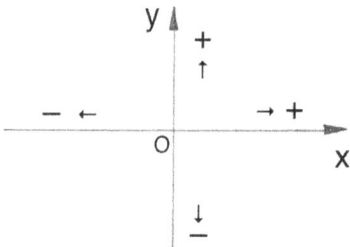

The choice of which direction gets to be called positive is arbitrary. Most engineers find it convenient, when dealing with an x-y reference frame, to use the positive sign to denote 'to the right' in the x-direction, and 'upward' in the y-direction.

Accordingly, a negative sign will denote 'to the left' in the x-direction, or 'downwards' in the y-direction.

If an object is to the left of the reference point, we say its displacement is negative. If it is to the right of that point, we say its displacement is positive.

Irrespective of its displacement, if an object is moving to the left, its velocity is negative, and, if it is moving to the right, its velocity is positive.

Likewise, irrespective of where an object is, or how fast it is going, if the rate of change of its velocity is increasing to the left, the object is experiencing a negative acceleration. If the rate of change of its velocity is increasing to the right, it has a positive acceleration.

It is quite possible that the displacement, velocity and acceleration of an object could have independently *different* signs at any given time.

Consider what happens when you are travelling in a car approaching a stop-street and brakes are applied. If we consider the direction in which you are moving as 'positive':

While applying brakes, you are still moving in the positive direction, but your velocity in the direction in which you are moving is diminishing, so we say you are decelerating, or experiencing an acceleration opposite to the one in which you are moving.

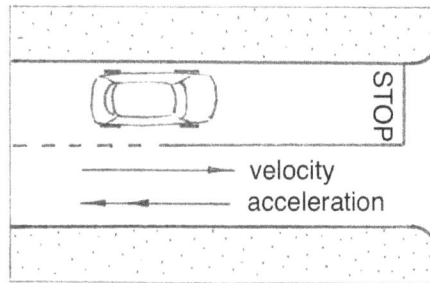

To designate this 'oppositeness', we say you are accelerating in the negative direction.

When solving problems in linear motion, choose a sign convention, and stick to it. Later, when you start using the the equations of motion, the mathematics will indicate the directions of any vector quantities that you need to determine, by your answers turning out respectively either positive or negative.

The elementary case of linear motion with constant speed

An object moving with constant speed will cover a certain amount of distance in a given time, according to the fundamental definition:

Speed = (distance / time) or, re-arranging this relation: **distance = speed × time**

The units of the LHS and the RHS of this equation need to be consistent. If the speed is given in m/s then all distances must be given in metres and all times in seconds. Likewise, if the speed is measured in km/h, distances must be given in km and times in hours.

So, if you are planning a trip in a light plane capable of flying at 200 km/h and the fuel will last you approximately 2 hours at cruising speed, you can bank on travelling approximately 400 km on one tank of fuel.

Illustrating the relation 'distance = speed × time' by means of a graph.

The following graph shows speed vs. time, for an object moving with a constant speed of 20 m/s:

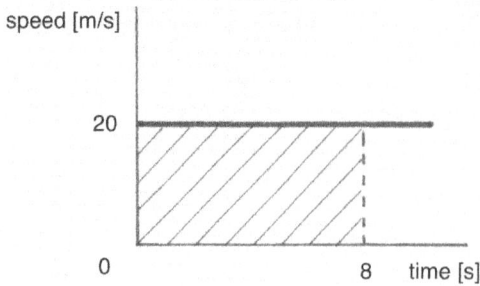

To determine how many metres this object moved in 8 seconds, we apply the relation: distance = speed × time, giving us:

distance = 20 m/s × 8 s = 160m.

However, note that the area under the graph between t = 0 and t = 8 seconds is a rectangle.

The value of that area is found by the product of the lengths of its two adjacent sides, namely speed (20 m/s) and time (8 seconds). The value of the area under this graph is thus equivalent to the distance covered between 0 and 8 seconds.

This illustrates the general observation that the area under a speed vs. time graph represents the distance covered.

Example

An athlete in a walking race passes an observer, striding at a constant speed of 3 m/s, and ten seconds later he suddenly slackens his pace to 2 m/s, which he keeps up for a further 15 seconds, before stopping abruptly. Assuming the speed changes are instantaneous, draw a graph of speed vs. time for this motion. How far from the observer would the athlete be after coming to rest?

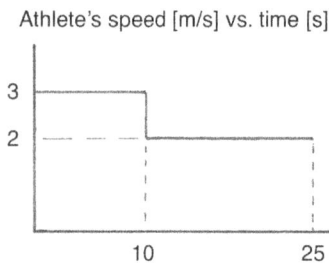

The distance covered is represented by the area under the graph. This consists of two rectangles, so the total area under the graph is

(3 × 10) + (2 × 15) = 60 m which is how far the athlete would be from the observer by the time he stopped.

Exercise

Consider the following motion of a particle confined to moving in one direction in a straight line, shown on a speed vs. time graph:

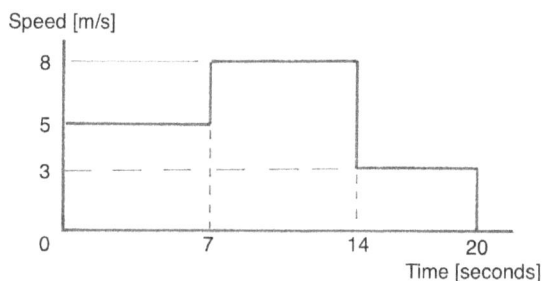

Between time t = 0 and t = 20, how far has the particle moved? [109 m]

The cases discussed above dealt with speed vs. time. When we expand our focus from 'speed' to 'velocity' (namely speed in a specified direction) we find that the area under the graph of velocity vs. time similarly represents *displacement*.

6

Before we can demonstrate this, we need to account for the fact that velocity changes do not occur instantaneously. The realistic motion of any object cannot be described by simple rectangular areas on a velocity-time graph. We have to account for periods of acceleration, during which there is a gradual change of velocity.

Acceleration and deceleration on a velocity-time graph

A positive acceleration implies an increase of velocity with time, which could be indicated on such a graph by any of the following lines:

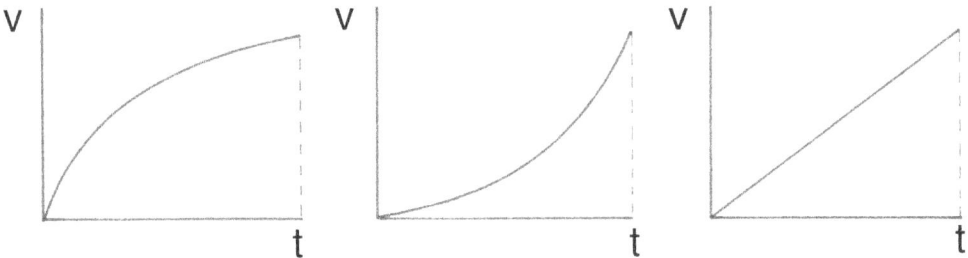

In the first diagram above, the value of the acceleration starts off high, and diminishes with time. In the second diagram, the acceleration begins gradually, and increases with time. In the third diagram, the value of the acceleration remains uniform over time.

Clearly, the value of the acceleration is represented by the gradient of the graph. The steeper the graph, the greater is the rate of change of velocity, namely the value of the acceleration.

Examine the units of the quantities represented on such a graph: gradient = (vertical) ÷ (horizontal), so the units of the gradient are (units of the vertical axis) ÷ (units of the horizontal axis), namely:

$[m/s] \div [s] = [m/s^2]$

For example, an acceleration of 4 m/s^2 indicates that the velocity is increasing by 4 m/s, every second.

This would be represented on a velocity-time graph as shown here.

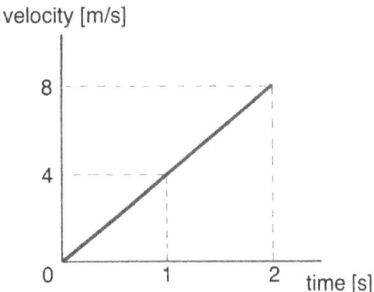

The cumulative displacement undergone by an object moving in a straight line path is represented by the area under the velocity-time graph.

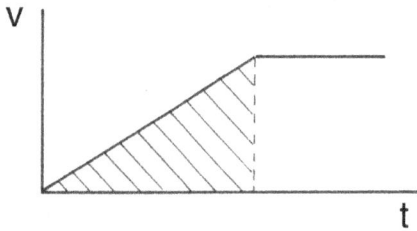

This applies whether the object is moving with constant velocity, or accelerating. If the object moves only in the positive direction, this cumulative displacement is identical with the distance covered. Therefore, if an object accelerates uniformly up to a given speed, the distance covered during that period of acceleration is represented by the area of the triangle under the graph.

Example

A train moves out of the station, from rest, with uniform acceleration 0.5 m/s² until it reaches its cruising speed of 72 km/h. How long does it take to reach this speed? How far from its starting point does it reach this speed?

Draw a velocity-time graph for this motion, noting that 72 km/h is equivalent to 20 m/s.

It is easier and advisable to work with the basic units of [m], [s], and [m/s] rather than to work with [km], [h] and [km/h]. Let the time taken to reach cruising speed be 't'.

The acceleration is represented by the gradient of the graph.

Gradient = (vertical) ÷ (horizontal)
∴ 0.5 = 20 ÷ t   ∴ t = 40 seconds

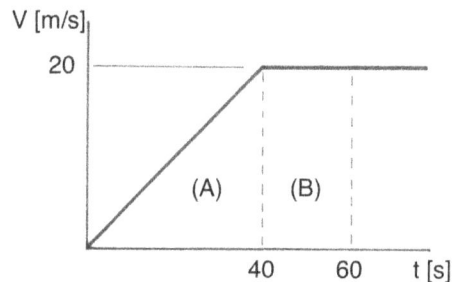

The distance covered while accelerating is represented by the area (A) under the graph, namely ½ (base) × (height) = ½ (40 × 20) = 400 m

What distance is covered in the first minute of the journey? Refer to the graph: this distance is represented by the sum of the areas (A) and (B), which amounts to 800 m.

Representing deceleration on the velocity-time graph

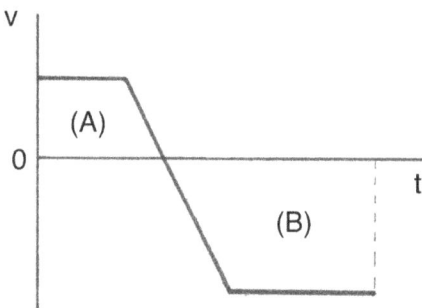

Since deceleration is associated with a decrease of velocity over time, it is represented on a velocity-time graph by a line or curve that drops to the right.

As with accelerating, the area under a portion of the graph that descends to the right represents additional displacement incurred.

(A) = displacement in the positive direction
(B) = displacement in the negative direction

8

Exercise

A train on a track accelerates at 0.4 m/s² for one minute, then continues at constant speed for a further minute, after which it decelerates at 0.5 m/s² until it comes to a stop. Sketch a velocity-time graph of this motion, and determine how far the train travels altogether.   [2736 m]

Interpreting velocity-time (v-t) graphs

Exercise

Match each one of the following labelled **v-t** graphs to one of the kinds of motion described below the diagram, by filling in the appropriate letter on the broken line next to the description:

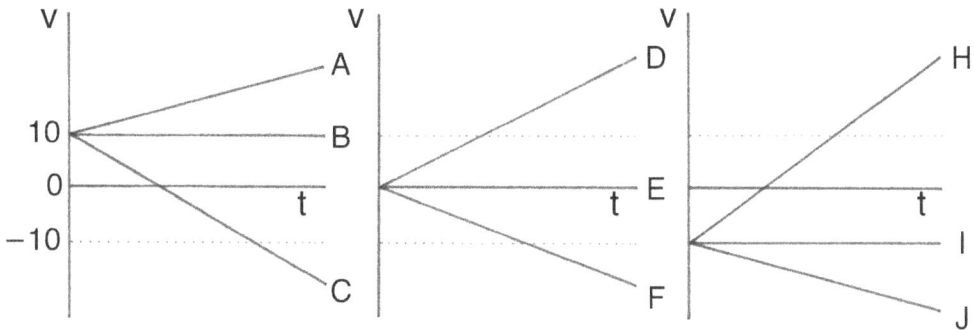

For example: Moving at constant velocity of 10 m/s... B

1. Starting at rest, then accelerating uniformly to the left ....
2. Moving to the left at **t** = 0, with constant velocity of 10 m/s ....
3. At **t** = 0, moving to the right, but decelerating uniformly ....
4. Moving to the left at t = 0, and accelerating to the left ....
5. Passing the observer at **t** = 0, while moving to the right and accelerating to the right..
6. Remaining at rest for the duration of the observation...
7. Moving to the left, at **t** = 0, but slowing down, eventually coming to a stop, changing direction, and picking up speed. ....

Exercise

Draw, on the set of axes given below, three **v-t** graphs, to represent each of the following motions of a vehicle, and label these graphs (K), (L), and (M) respectively:

*   At **t** = 0, moving to the left at 8 m/s, but slowing down with uniform deceleration, until its velocity is 15 m/s to the right (having come to rest momentarily at **t** = 10).

9

- At $t = 0$, starting from rest, accelerating uniformly over 8 sec to a velocity of 20 m/s to the right, then decelerating uniformly to rest over the next 120 m.

- Passing the observer at $t = 0$, while moving at a constant velocity of 24 m/s to the right, then, after 12 seconds, decelerating at 3 m/s² for 4 seconds, after which, continuing at constant velocity.

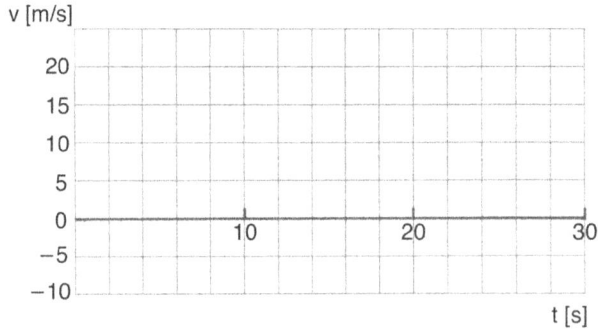

Exercise

On a straight stretch of road, a motor-car passes an observer from left to right, doing a steady 108 km/h. Six seconds later, the driver sees a speed limit sign indicating a reduced speed zone. He applies brakes, dropping his speed uniformly to 72 km/h over a distance of 200 m. He holds this speed steadily for a further 20 seconds, before braking again, this time to stop outside a cafe. In braking to come to a stop, he decelerates at a rate of 4 m/s².

a. Draw a v-t graph to scale, to represent this series of motions.

b. Determine the time taken respectively for each period of deceleration. [8 sec; 5 sec]

c. What is the distance between the observer and the cafe?   [830 m]

d. What is the average speed of the car for the stretch between the observer and the cafe?   [78.63 km/h]

Exercise

Consider the following v-t graph for a given series of movements by a slider on a track.

Determine:

The maximum *value* of the acceleration at any time. [3 m/s²]

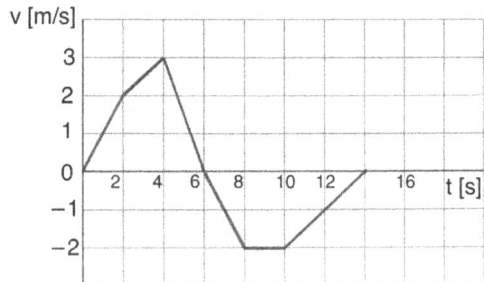

The total distance moved between time
$t = 2$ and $t = 8$ seconds   [10 m], and

The displacement from the starting point, when $t = 10$ seconds.   [4 m]

Exercise

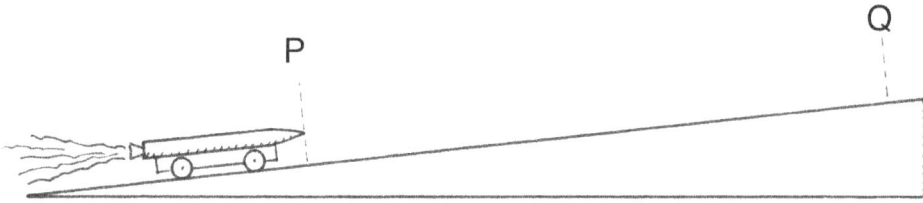

On a sloping stretch of straight track, a rocket-powered trolley, initially at rest at point P, moves uphill with uniform acceleration of 4 m/s² for five seconds. The motor is cut, and the trolley continues freewheeling until eventually, 25 seconds after starting, it has slowed to a stop at point Q, and starts to roll backwards until it passes point P on its way down the track.

a. Draw a v-t graph to represent this series of motions.
b. What is the maximum velocity reached by the trolley before beginning to slow down?   [20 m/s]
c. What is the distance between point P and point Q?   [350 m]
d. How long does the whole process take from starting at point P, to passing point P again on the way down?   [59.58 sec]
e. What is the speed of the trolley when it passes again through point P?
   [– 23.66 m/s]

Derivation and application of the three basic equations of linear motion with uniform acceleration

Consider an object that is confined to move on the path of a straight line.

Suppose the object accelerates with uniform acceleration 'a', from an initial velocity 'u' to a subsequent velocity 'v', over a time interval of 't' seconds.

The acceleration, 'a' is defined to be the rate of change of velocity with time:

$\therefore\ a = (v - u) \div t$

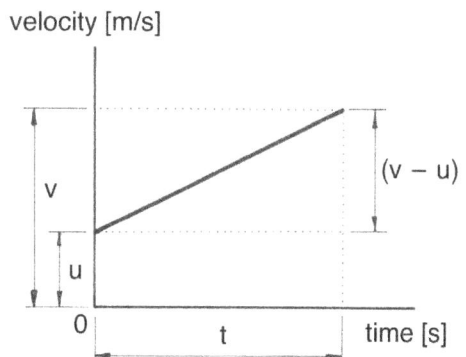

Rearranging this equation, we get:

v = u + at ....(A)

Now, displacement s = (average velocity) × time, where the average velocity during time 't' is (u + v)/ 2

Therefore the displacement, s, is given by: s = ((u + v)/2) × t ....(i)

Substituting the value for 'v' from equation (A) into (i):

s = ½[u + u + at]×t

∴ **s = ut + ½ at²** ....(B)

This equation can be verified by examining the areas under the velocity-time graph above:

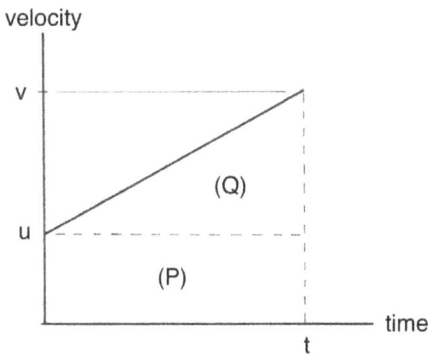

Area (P) = base × height = t × u = ut

Area (Q) = ½ base × height

$$= ½ (t) (at)$$

$$= ½ at^2$$

Hence, the total area under the graph (which represents displacement) is equal to ut + ½ at², thus verifying equation (B)

A third useful equation is found by eliminating the variable 't' :

From (A):   t = (v − u)/a

Substitute this value into equation (i):   s = [(u + v)/2] [(v − u)/a]

∴ 2as = (u + v)(v − u) = v² − u², hence   **v² = u² + 2as** ....(C)

Any exercise involving linear motion with uniform acceleration may be solved either by the correct application of equations (A), (B) and (C), or by determining the areas and gradients on a velocity-time graph, by simple geometry, as we did above, before we derived these equations.

The correct application of these equations requires that initial conditions must be specified for *each* of the phases of the linear motion.

In the motion described by the following graph, for example: there are four phases of motion, labelled respectively: (1), (2), (3) and (4).

The equations have to be applied to each phase separately. They cannot be applied in an

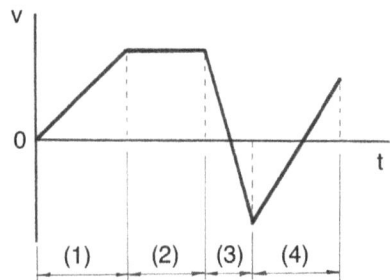

overall sense to the entire motion, because the value of the acceleration changes from one phase to the next.

Example

A vehicle is climbing a long straight uphill road. At time $t = 0$, the vehicle passes an observer at a speed of 15 m/s, accelerating at a uniform rate of 3 m/s^2. Ten seconds later, the engine breaks down, and the vehicle coasts to a halt, decelerating at 1 m/s^2 all the way. The driver does not apply brakes until the vehicle stops of its own accord. Using the appropriate equations of motion, determine how far from the observer the vehicle came to a halt.

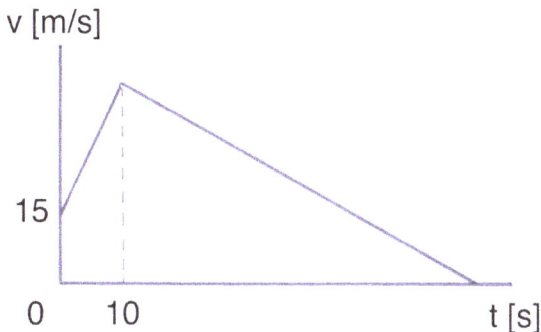

First, draw a v-t graph of the motion. There are two phases of motion here, firstly the acceleration phase, and then the deceleration phase.

Phase 1: Initial velocity $u = 15$m/s; acceleration $a = + 3$ m/s^2.

We need to determine the value of displacement **s** when $t = 10$ sec.

From $s = ut + \frac{1}{2} at^2$: $s = 15(10) + \frac{1}{2} (3)(10)^2 = 300$ m

Phase 2: We need to know the initial velocity for this phase: This initial velocity is the same as the final velocity for phase 1, so we apply the equation:

$v = u + at$ *to phase 1*, yielding $v = 15 + 3(10) = 45$ m/s

Hence, for phase 2, initial velocity, $u = 45$ m/s. We also know that the final velocity for phase 2 is zero, as the vehicle slows to a halt, and the acceleration value is -1 m/s^2.

For phase 2: use the equation $v^2 = u^2 + 2as$ Thus: $s = (v^2 - u^2) / 2a$

$\therefore s = (0^2 - 45^2)/(2)(-1) = 1012.5$ m

The total distance covered is equal to the sum of the displacements for phase 1 and for phase 2, namely

300 m + 1012.5 m = 1312.5 m.

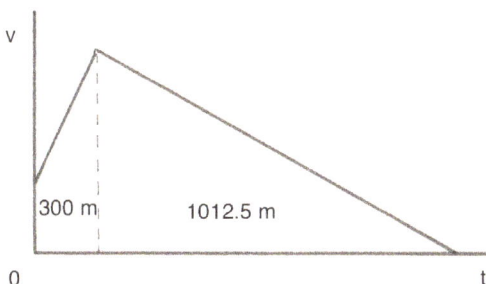

Exercises on linear motion, set 1

1. A car moving at 72 km/h decelerates uniformly to stop in a distance of 100 m. Determine the value of the deceleration. [2 m/s^2]

2. If a vehicle moving at 108 km/h decelerates at 2 m/s^2, what will be the braking distance? [225 m]

3. A sprinter accelerates from rest, to reach his maximum speed of 10 m/s, over a distance of 25 m. If this acceleration is uniform, how much time does it take to reach maximum speed, and what will be his time for the 100 m race? [5 sec; 12.5 sec]

4. If you are in a car, travelling at 120 km/h a distance of 50 m behind another car, and the car in front experiences some sudden disaster (such as being hit by an oncoming car) can you stop in time to avoid colliding with it, if you hit the brakes, applying your maximum deceleration of 4 m/s^2? [no, it will take 139 m for you to stop]

5. Two trains on parallel straight tracks start from a terminus at the same time, moving in the same direction. Train A accelerates uniformly at 0.5 m/s^2 until it reaches a steady velocity of 72 km/h, which it maintains for some time. Train B does likewise, but its acceleration is 0.3 m/s^2 and its cruising speed is 108 km/h. How far apart are the trains, and which one is ahead, after 1 minute, and 4 minutes, respectively? [A: by 260 m; B: by 1300 m]

Exercises on linear motion, set 2

Question 1

This velocity-time graph illustrates the motion of two cars, A and B, on a straight, level stretch of road, both moving in the same direction. The cars are level with each other at time t = 80 seconds.

Using the graph, determine:

Which one was ahead, and how far, at time t = 0? [B was 100 m ahead]

If they both apply brakes at time t = 80 seconds, and both decelerate uniformly at the same uniform rate of 2 m/s^2 until they stop, then, which one will be ahead, and by how far, when they have stopped? [A will be 225 m ahead]

How many seconds after the first car has stopped will the second car stop? [5 sec]

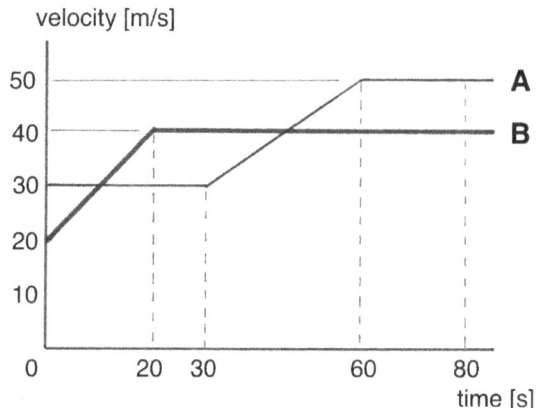

velocity [m/s]

time [s]

Question 2

Assume that a motor-car accelerates uniformly in each gear. The following velocity-time graph depicts the motion for a short journey by this car on a straight level road. The car accelerates through three gears, continues at constant velocity for a while, then brakes with uniform deceleration until it comes to a stop.

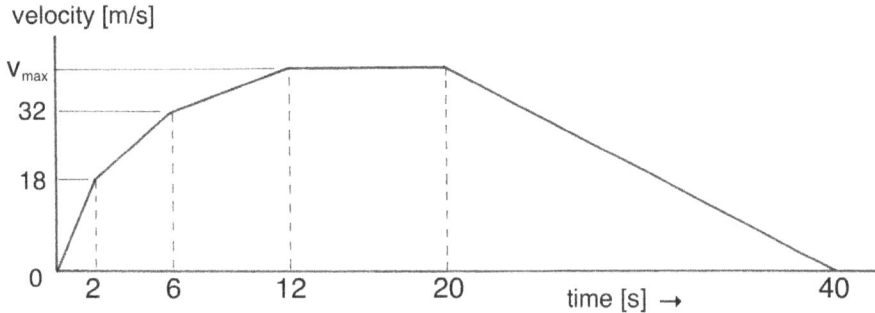

If it travelled at constant velocity for 320 m, what was the value of this velocity? [40 m/s]

What was the braking distance? [400 m]

How far did it travel altogether? [1054 m]

Question 3

If a car has a maximum speed of 144 km/h, a maximum uniform acceleration of 2.5 m/s^2 and a maximum uniform deceleration of 4 m/s^2, what is the furthest distance it can travel in a straight line, from starting to stopping:

a. In one minute? [1880 m]

b. In 20 seconds? [307.7 m]

Exercises on linear motion, set 3 *(Suitable for solving in groups)*

Question 1

Two go-karts race on a straight track, from a standing start. Kart A has 3 forward gears, and kart B has four.

The table below shows the magnitudes of the accelerations in each gear (in units of m/s^2), and the time for which each vehicle is in each gear (in seconds).

Kart	Accel. in 1st gear	Time in 1st gear	Accel. in 2nd	Time in 2nd	Accel. in 3rd	Time in 3rd	Accel. in 4th	Time in 4th
A	1.2	5	0.5	10	0	ongoing	N/a	N/a
B	3.0	1.5	1.0	3	0.4	5	0	ongoing

15

Draw an accurate velocity-time graph to scale, showing the two motions. *Use the scales: 10 mm = 1 second and 10 mm = 1 m/s.*

Determine, with the assistance of the graph: which car is ahead, and by how many metres, 17 seconds after the start of the race.   [B is ahead, by 13.13 m]

If the race is over a distance of 220 m, what will be the winning time for the race and which car will win?   ['A' wins, with a time of 25.91 sec]

Question 2

A particle is confined to move along a straight line. At time t = 0, it begins moving from rest at point P, and accelerates to the right at 3 m/s^2 for 4 seconds. At t = 4, it accelerates to the left at 2 m/s^2 for a further 14 seconds, after which it continues with constant velocity for another 4 seconds. Determine:

a.   The maximum velocity to the right at any time in this sequence.   [12 m/s]

b.   The maximum velocity to the left at any time in this sequence.   [16 m/s]

c.   The maximum displacement to the right of point P.   [60 m]

d.   Where the particle is at t = 22 seconds.   [68 m to the left of point P]

e.   The total distance covered in this sequence of movements.   [188 m]

Making use of a displacement-time graph

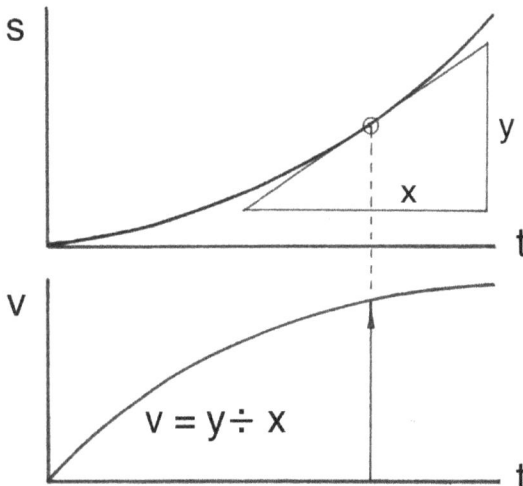

Thus far, we have been working with the velocity-time graph for linear motion.

Now we introduce the *displacement-time* graph, which can be useful in circumstances in which it is difficult to measure *velocity* at specific times, but possible to measure *displacements*.

We know that acceleration, being the rate of change of velocity with respect to time, is represented by the gradient of a velocity-time graph.   **a = dv/dt**

Similarly, since velocity is the rate of change of *displacement* with time: velocity is represented by the *gradient* of a displacement-time graph. **v = ds/dt**

The instantaneous velocity at each of the points shown: $v_1 = y_1 \div x_1$

and so on.

If you are able to plot a displacement-time graph for a movement observed in reality, you can use the graph to estimate the velocities attained by the moving object, from the gradient of the graph at various points.

These velocity values can be used to plot a second graph, namely a velocity-time graph, for the motion that was observed. The v-t graph, in turn, can be used to estimate the values of the acceleration at specific time intervals, as in the following experiment:

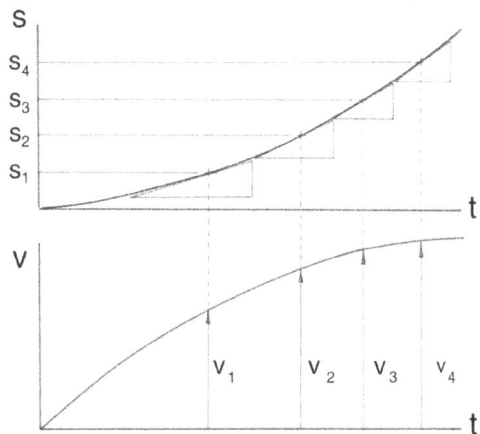

A practical experiment to compare the initial accelerations of two motorcycles

The following practical activity has been carried out by students of the author. The students wanted to see which of two makes of motorcycle had the greater initial acceleration.

Aim: to estimate the value of the initial acceleration of a motorcycle accelerating from rest down a straight stretch of road.

Method: The experiment was done on a straight stretch of tarred road that was closed to the public. A line of observers with stopwatches was posted alongside this road at 20 m intervals, for a distance of 300 metres.

A motorcyclist took off from rest at the starting point, and accelerated down the road until he was beyond the last observer.

The signal for him to start was the waving of a flag, which all the observers could see.

On seeing the flag go down, each observer started his stopwatch. Each respective observer measured the time elapsed from the start until the motorcyclist was level with him.

The observations enabled a plot of displacement vs. time to be made. The gradient of the graph was then estimated at suitable time-intervals.

The estimated gradient values, indicating velocities, were then plotted to construct a velocity-time graph.

A similar procedure was applied to estimate the accelerations, from measuring the gradient at various points on the velocity-time graph. In this way, the initial accelerations of two motorcycles could be estimated and compared.

Exercises on linear motion, set 4

Question 1

A motoring enthusiast takes his new car for a few practice runs, accelerating as fast as possible from a standing start, on a straight, level road. With the help of a friend, he records the following data, as averaged over a few runs:

Gear	1st	2nd	3rd	4th	5th
Max vel reached in this gear [km/h]	60	110	160	190	200
Time in this gear to reach this velocity [s]	2	4	6	8	10

a. Draw to scale a velocity-time graph for the motion of this car, using [m/s] as the units of velocity. Assume all accelerations are uniform.

b. What is the magnitude of his average acceleration in first gear?   [8.33 m/s^2]

c. Determine the distance covered by the time the driver changes to 5th gear. [725 m]

d. Determine the shortest time he can cover 1 km from a standing start. [35.15 seconds]

Question 2

A motorcyclist takes off from a standing start and accelerates as quickly as he can past a row of observers stationed in a straight line, at intervals of 30 m. Each observer records the time taken from the start to when the motorcyclist is level with him. The data they collect is as follows:

Observer	1	2	3	4	5	6	7	8	9	10
Time from start [sec]	2.10	3.10	3.88	4.74	5.52	6.28	7.02	7.73	8.43	9.13

a. Use this data to draw in the first instance a *displacement vs. time* graph, to scale, for the motion of the motorcycle.

b. Draw a second graph, directly below the first graph, this one a *velocity - time* graph for the motion, to the same time scale as the one above it.

From the graphs, *estimate* (note, these will be *estimates*, depending on the accuracy of your constructions):

The velocity at t = 1 [approx. 48 km/h]

The velocity at t = 4 [approx.129 km/h]

The maximum velocity reached [approx. 154 km/h],

The average acceleration between t = 0 and t = 1. [approx.13.5 m/s^2]

Test your understanding of linear motion with uniform acceleration

The true/false test below may be used, either individually, or in discussion with other members of a small group of students, to check how well you have understood the principles explained in the present chapter

True/false test on linear motion with uniform acceleration

	Circle either 'true' or 'false' next to each statement		
1	Velocity and speed are always different.	T	F
2	Displacement can be equal to distance covered, in some instances.	T	F
3	If the direction of motion of an object is changing, its velocity must be changing.	T	F
4	Accelerations to the right must necessarily have a positive sign.	T	F
5	If the acceleration of an object is negative, it means the object is slowing down.	T	F
6	If a velocity-time graph slopes downward to the right, it means the acceleration is negative.	T	F
7	If a velocity-time graph slopes downward to the right, it means the velocity is negative.	T	F
8	The area between a velocity-time graph and the t-axis, from time 1 to time 2, represents the cumulative displacement between those two times.	T	F
9	For an object moving with constant speed, time = distance ÷ speed	T	F
10	A motor vehicle is capable of greater acceleration in 4th gear than in 1st gear.	T	F

11	On a velocity-time graph, if the total area between the graph and the time-axis above the time-axis is equal to the total area below that axis, it means that the object whose motion is being described is momentarily at rest.	T	F
12	If the whole of a velocity-time graph is below the t-axis, it means that all displacement undergone has been in the negative direction.	T	F
13	The position of an object on its line of movement cannot be read directly off a velocity-time graph.	T	F
14	The basic equations for linear motion can only be applied for periods of time for which the initial and final velocities are known to begin with.	T	F
15	Assuming that when brakes are applied, a car is capable of a maximum deceleration of 4 m/s^2, then the braking distance required to come to rest from a speed of 30 m/s is four times that needed to come to rest from 15 m/s.	T	F
16	The gradient of a displacement-time graph represents acceleration.	T	F
17	On a velocity-time graph, any area that lies below the time-axis, represents events that took place before time zero.	T	F

Chapter 11

Motion influenced by gravity: vertical motion and projectile motion

How gravity affects vertical motion

Solving vertical motion exercises using the three equations of linear motion

Dealing with x- and y- motions separately

Analysing the trajectory of a projectile

A practical way of determining the launch velocity of a high speed projectile

The observed effect of air resistance on the trajectory of a golf ball

The first part of the chapter applies to objects dropped, or given an initial vertical velocity, either upward or downward, and then allowed to move freely, under the action of gravity. The second part of the chapter deals with the motion of projectiles.

How gravity affects vertical motion

The Earth's gravity causes all objects that have mass to be attracted toward the centre of the Earth, with a gravitational force that is a function of the mass of the object. Recall from an earlier chapter that the value of this force (namely, the weight of the object) is given by the equation expressing Newton's Law of Gravitation, namely:

$W = GMm/r^2$ (*Which equation is only valid provided that neither object is inside the boundary of the other.*)

Since the acceleration of a mass 'm' acted on by a force 'F' is given by:
$a = F/m$ (Newton's second law), it follows that the gravitational acceleration
$g = force \div mass = W/m$ $\therefore g = (GMm/r^2) \div m = GM/r^2$

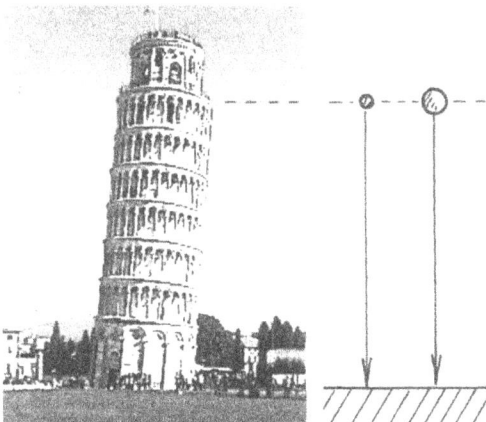

Which shows that the value of g is *independent* of the value of 'm'. So, for a given value of 'r', the acceleration due to Earth's gravity that any object undergoes will have *the same value,* irrespective of its mass.

Galileo illustrated this truth in a famous demonstration, by simultaneously dropping a heavy object alongside a lighter object, from the leaning tower of Pisa.

The two objects hit the ground at the same time.

This result surprised a lot of people who had believed that heavy objects must fall faster than lighter ones. You can repeat this experiment from any convenient high position, provided you make sure that the air resistance on the two objects is not significantly different.

Reminder: we saw in the chapter on forces why the value of 'g' differs slightly from place to place on Earth. For most practical purposes in engineering, the generally accepted average value of 9.81 m/s^2 may be used. The actual value of 'g' can vary marginally from this value. If you need a high degree of precision for some application, you can take the trouble to find out the exact local value.

Variations in the value of 'r': Note that the highest point on Earth (Mt. Everest) is around 8.8 km above sea level. The deepest mine in the world is around 3.9 km deep at the time of writing. Compared with an average Earth radius of 6371 km, the variation in 'r', from one place to another in the general region of engineering activity, is negligible.

When does gravitational acceleration affect motion?

Gravity is present *all the time*. It provides a downward force, which is continuously present, and, if not opposed, will result in an object accelerating towards the centre of the Earth at 9.81 m/s^2, unless the moving object:

- Is experiencing another force simultaneously (such as air resistance or buoyancy) or
- Has come to rest against an obstacle (such as hitting the ground), or
- Is a significant distance from the surface of the Earth, in which case the value of g will differ, according to that distance.

If an object is restrained by some other force, such that the force of gravity is being partially opposed, then the actual downward acceleration of the object will be diminished (as when the object is falling through water).

When falling through air, or through water, the resistance depends on the drag coefficient of the object, which is a function of its shape and size. (See more about drag coefficients in the chapter on vehicle dynamics, in volume 3 of this series.)

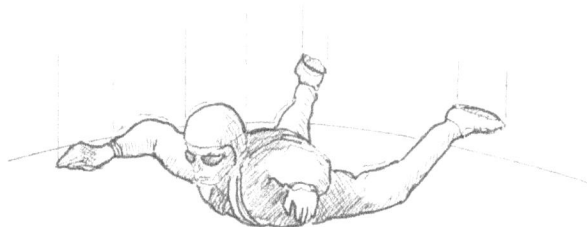

The magnitude of the force due to air resistance increases with the speed of the object, so, if an object has fallen for a long enough distance, it is possible for it to reach a velocity at which the downward force due to gravity is equal to the upward force due to air resistance.

The velocity at which this occurs is called the terminal velocity, because, after reaching this velocity, the object can no longer accelerate.

An object that has reached its terminal velocity while falling in air, will maintain that velocity until it is acted on by some other force. Sky-divers in free fall usually reach terminal velocity some while before they activate the parachute.

For the purposes of the analysis provided in this chapter, we will assume air resistance to be negligible. This assumption is to some extent difficult to justify, since air resistance is far from negligible, and is actually very significant at high velocities. However, for the present purpose, we can justify the assumption, since:

The velocities encountered in the examples used here will be relatively low, and

The objects chosen for our examples have small drag coefficients (stones, balls of high density and missiles, like arrows)

Once the principles of solving vertical motion exercises have been learnt, the reader may wish to introduce some realistic parameters that more closely replicate the effects of air resistance.

Solving vertical motion exercises using the three equations of linear motion

Choosing a sign convention: Vertical motion is merely a special case of linear motion with uniform acceleration, in which the value of the acceleration is equal to 'g'. While it is common practice to replace the acceleration symbol 'a' with 'g' in such exercises, it is not essential to do so. In this chapter we will continue to use the three equations of motion in their standard form, namely

$v = u + at$; $s = ut + \frac{1}{2}at^2$ and $v^2 = u^2 + 2as$.

We will apply the convention that the upward direction is positive, and downward is negative. The value of the acceleration in our equations will therefore always be: $a = (-9.81 \text{ m/s}^2)$

With this sign convention, positive values of displacement will indicate position *above* the selected reference level, and negative values will indicate how far *below* the reference level the moving object is.

Positive values of velocity will indicate movement in the upward direction, and so on.

When solving a problem involving vertical motion, it is highly advisable to make a sketch of the physical flight path of the object, showing the reference level for

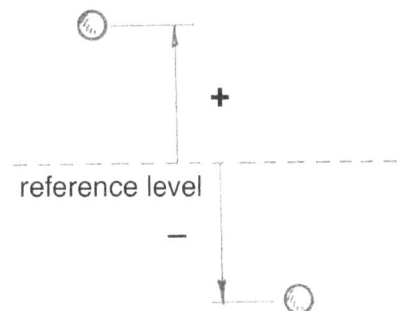

vertical displacement, which is usually chosen to coincide with the point from which the object is launched or dropped. It is useful also to show the turning point at the top of the trajectory, for objects that are projected upwards.

Example

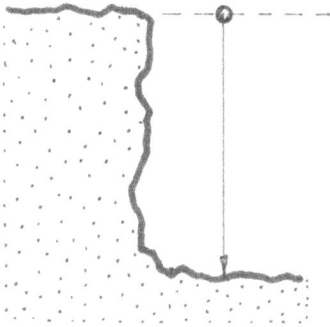

A stone dropped from a cliff-top takes 3.6 seconds to hit the ground below the cliff. How high is the cliff?

Draw a sketch of what is physically happening.

Specify the reference level and all initial conditions: initial velocity u = 0, and acceleration a = – 9.81 m/s².

Since a value for 't' has been given, and a displacement is needed, use the equation:

$s = ut + ½at²$ ∴ $s = 0(t) + ½(– 9.81) (3.6)²$
= – 63.57 m.

The negative sign of this answer indicates that, at t = 3.6 seconds, the displacement of the stone is 63.57 m *below* the point from which the stone was released. This means the cliff is 63.57 m high.

The v-t graph for the motion of the stone looks as follows:

A further question: With what velocity does the stone hit the ground?

Using v = u + at: v = 0 + (– 9.81)(3.6)

= – 35.32 m/s

The negative answer indicates that the velocity is downward at the time of colliding with the ground.

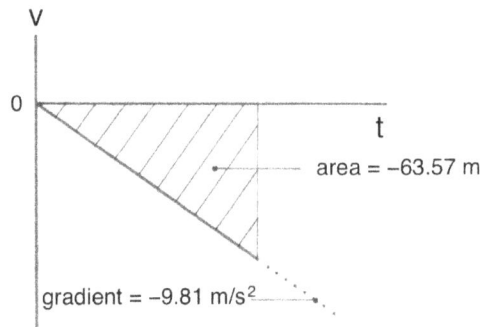

area = –63.57 m

gradient = –9.81 m/s²

Example

If you throw a ball vertically upwards,
and it takes 5 seconds from leaving your hand to fall back to the level where it left your hand:

With what velocity did you throw it?

How long does it take to reach the top of its path before falling back?

How high above the point of projection does it go?

Let the reference level for displacement be the point of projection. Let the initial velocity of the ball be u. When the ball is again at this level, s = 0.

Use this fact in the equation:

$s = ut + \frac{1}{2}at^2$

$\therefore 0 = u(5) + \frac{1}{2}(-9.81)(5)^2$

$\therefore u = 24.53$ m/s The positive sign confirms that the initial velocity was upwards.

At the turning point at the top of the flight path of the ball, v = 0. Use this fact in the equation

$v = u + at$:

$\therefore 0 = 24.53 + (-9.81)t \quad \therefore t = 2.500$ seconds

Notice that this time is exactly half the time it took for the whole flight of the ball.

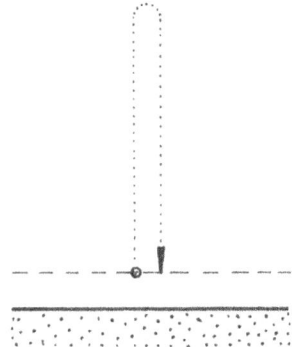

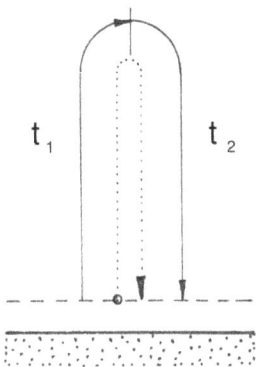

In all exercises of this type, the time taken to reach the turning point is exactly half of the total time taken from leaving the reference level to returning to the reference level. Namely, time $t_1 = t_2$.

To determine the height at the top of the flight path: now that we have the value of the initial velocity, as well as the time taken to get to that point, we can use these values in the equation: $s = ut + \frac{1}{2}at^2$

$\therefore s = 24.53(2.5) + \frac{1}{2}(-9.81)(2.5)^2 = 30.66$ m

The v-t graph for this motion looks as follows:

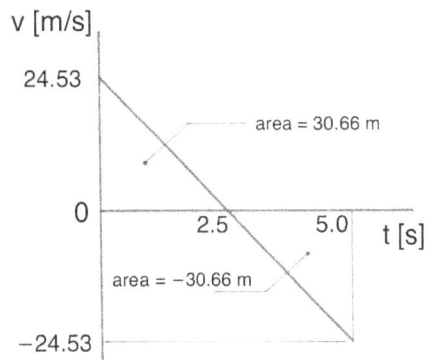

25

Example

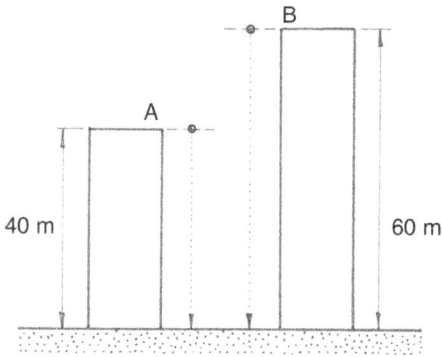

Two points, A and B, are on the rooftops of adjacent buildings. A stone is dropped from point A, and at the same instant, another stone is thrown vertically downwards from point B.

If the two stones reach the ground at the same time, what was the initial velocity of the stone thrown from point B?

For stone A: u = 0, and when it hits the ground, s = – 40

Use these facts in the equation $s = ut + \frac{1}{2}at^2$: $– 40 = 0(t) + \frac{1}{2}(– 9.81) t^2$ ∴ t = 2.856 sec.

The two stones reached the ground at the same time, therefore the time for stone B to hit the ground was also 2.856 seconds. Using this data in $s = ut + \frac{1}{2}at^2$:

For stone B: $– 60 = u(2.856) + \frac{1}{2}(– 9.81)(2.856)^2$

∴ u = – 7.000 m/s (namely, 7.000 m/s downwards)

The v-t graphs for the two motions look as follows:

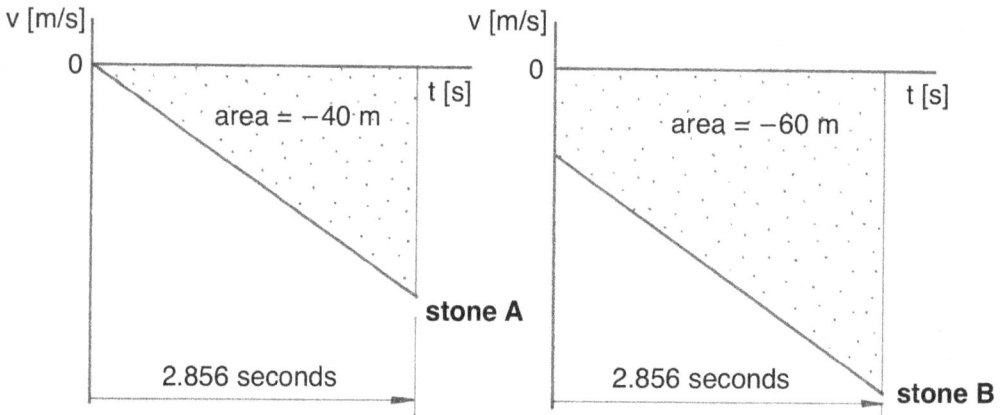

Exercises on vertical motion, set 1

Question 1

A ball thrown vertically upwards from the ground takes 4.2 seconds to reach the ground again. What was its initial velocity?   [20.60 m/s]

26

Question 2

An arrow is shot vertically upwards from the top of a 50 m high cliff. The arrow reaches a height of 150 m above the cliff-top before dropping back. A slight breeze carries it just wide of the cliff-top, and it hits the ground at the foot of the cliff.

Determine the launching velocity, and the speed with which the arrow strikes the ground. [54.25 m/s ; 62.64 m/s]

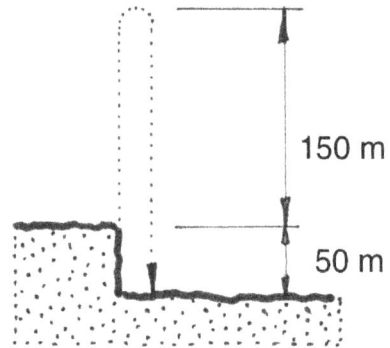

150 m

50 m

Question 3

A stone is shot vertically upwards from a catapult, with an initial velocity of 30 m/s. Determine the time taken to reach its turning point, and the maximum height reached.   [3.058 seconds; 45.87 m]

Question 4

Ball A is dropped from the top of a 30 m high building. At the same instant, another ball, B is thrown vertically upwards from the ground below the building. The two balls pass one another 20 m above the ground. What is the speed with which ball B was thrown?   [21.01 m/s]

Examples with a time delay between the launching of two projectiles

In the preceding examples, if there have been two objects on vertical paths of motion, they have both begun their motion at the same time. A complication arises when one of them is projected or released after the other. The correct approach to problems of this type will be demonstrated by means of an example.

Example

Ball A is thrown vertically upwards from the ground, with an initial velocity of 25 m/s. Four seconds later, Ball B is thrown upwards on a parallel path, with an initial velocity of 20m/s. How long after ball A is thrown will the two balls be level with each other? How long after ball A is projected will ball B hit the ground?

First sketch the v - t graph for ball A.

In order to make this sketch to scale, we need to know the time taken for ball A to get to the top of its path. At the top of the flight path, $v = 0$, so, using the equation

$v = u + at$, we get $0 = 25 + (-9.81)t$ and hence $t = 2.548$ seconds. Doubling that time (namely, 5.096 seconds) will give the time from launching to when ball A hits the ground. This gives us enough information to sketch the v-t graph for ball A.

Area 1 represents the positive displacement undergone by ball A, from being thrown, until it reaches its turning point.

Area 2 represents the negative displacement of ball A, undergone on its way down from its turning point, up to the moment when it hits the ground.

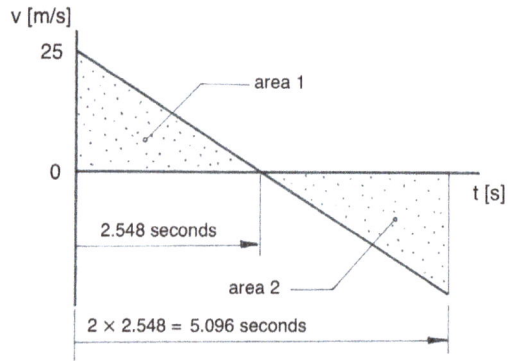

To apply the equations of motion to the movement of ball B, we note that the reference instant for time (namely, t = 0) occurs at the start of the motion of ball A. So, if ball B starts moving 4 seconds later, the time at the start of ball B's movement is t = 4 seconds.

So, when we set up equations to deal with the flight of ball B, should we use in place of (t) the expression (t + 4) or (t - 4) ? The reasoning is as follows:

For whatever duration that A has been in flight at any time, ball B has been in flight for 4 seconds *fewer*, and therefore the correct way to deal with the time variable in all equations relating to ball B is to use (t – 4) instead of (t). If this is done, all values of 't' for both projectiles will refer to the same time scale.

Since the criterion for solving the problem is that the two displacements are equal when the two balls are level with each other, we use the equation:

$s = ut + \frac{1}{2}at^2$ and, setting $s_A = s_B$:

$25t + \frac{1}{2}(-9.81)t^2 = 16(t - 4) + \frac{1}{2}(-9.81)(t - 4)^2$, which yields t = 4.629 seconds

The following v-t graph illustrates the movement of ball B, added to the original graph:

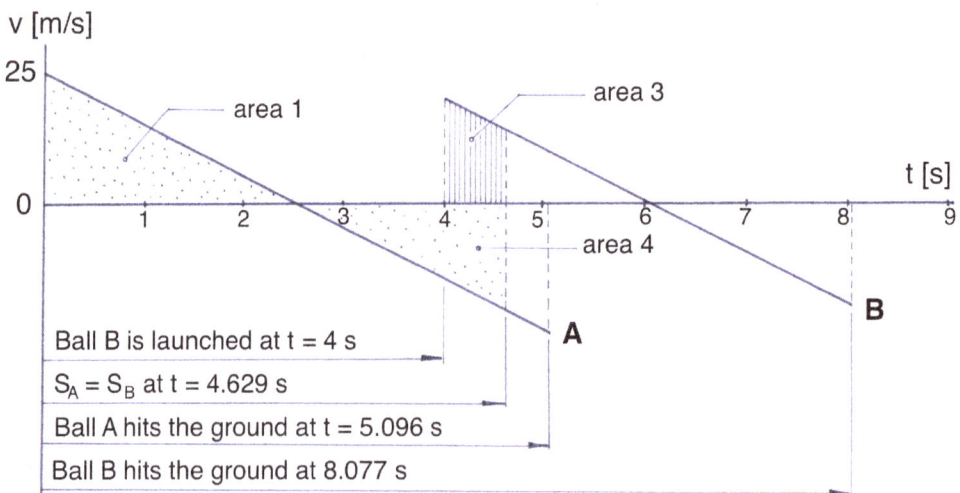

28

The graph above shows the following features of the respective motions of balls A and B:

- Since the displacements of the two balls are equal at the time they are level, the cumulative area under the two graphs at that moment must be equal. Hence, area (3) must be equal to area (1) minus area (4)

- At the moment the two balls were level, ball A was on its way down, while ball B was still ascending.

- The times that the two balls respectively hit the ground are indicated on the graph. The value of 8.077 seconds is obtained as follows: At the instant of hitting the ground, the displacement of ball B, s = 0. If this value is substituted into the equation for its displacement, namely $s = ut + \frac{1}{2}at^2$, we get: $0 = 20(t - 4) - 4.905(t - 4)^2$, which, when solved for t, yields two solutions, namely: t = 8.077 seconds and t = 4.000 seconds.

Exercises on vertical motion, set 2

Question 1.

A stone is projected vertically upwards from a catapult at the top of a 40 m high cliff, with initial velocity 50 m/s. Nine seconds later, another stone is dropped from the same position, to fall to the ground below the cliff-top. At what time, if any, are the two stones level?   [10.38 seconds after firing]

Question 2

At time t = 0, a sack of grain, P, is dropped from a building 120 m high. Three seconds later, an arrow, Q, is fired downwards from the same position, with an initial velocity of 60 m/s. Assuming the arrow was properly aimed, is it possible for it to hit the sack of grain before the sack reaches the ground?

[Yes, they are at the same level at t = 4.444 seconds, whereas the sack hits the ground at t = 4.946 seconds.]

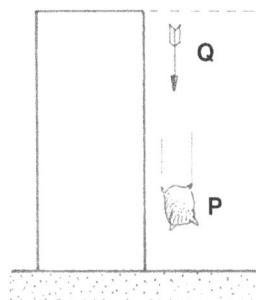

Question 3

A firework rocket, mass 120 g, is fired vertically upwards. The fuel lasts for 3 seconds, and while it is burning, it provides a propulsive force of 4 N. Ignoring the effects of air resistance and the diminishing mass of the remaining fuel:

    a. Sketch a v - t graph for the motion, and determine:

    b. The upward acceleration while under power   [23.52 m/s^2]

    c. The height reached under power [105.9 m]

d. The maximum height reached altogether   [359.7 m], and

e. The time from take-off to hitting the ground.   [18.76 seconds]

f. Now, re-draw the velocity-time graph to show the approximate effects of air resistance and diminishing fuel load. No calculations required.

Projectile motion

Most projectiles that we launch are not directed exactly vertically, but at some angle to the horizontal. Such projectiles include sports balls, stones, spears, arrows, cannon-balls and bullets. All of these are free to move under the influence of gravity after being launched.

Obviously, rockets and powered missiles do not fit in this category.

In the analysis that is used here, we will ignore the air resistance on a projectile, although it needs to be appreciated that air resistance does affect the flight of projectiles significantly, especially if their launching velocities are high. There is more about this at the end of the chapter.

Dealing with x- and y-direction motions separately

To analyse the motion of a projectile, it is convenient to choose an x-y co-ordinate system such that the x-axis is parallel to the Earth's surface, and the y-direction is aligned with the direction of the gravitational pull.

The reason for this choice is that the motion of a projectile has, at any given moment, two velocity components, one in the x-direction and one in the y-direction. It is important to realise that these two velocity components are governed by different rules:

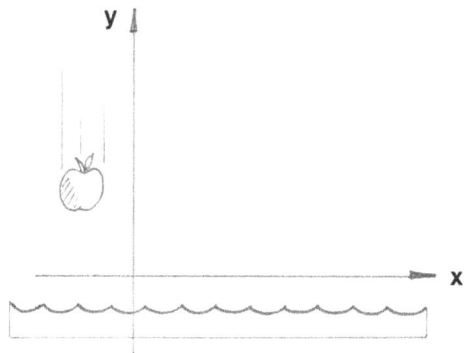

The motion in the x-direction is *not* affected by gravity. (Recall from the chapter on forces that a force can have no effect at right angles to itself.) Excluding the effects of air resistance, if a projectile is given an initial velocity in the x-direction, it will continue moving with that velocity in the x-direction, unless it experiences air resistance, or until it strikes an obstacle.

The motion in the y-direction *is* governed by gravity, in exactly the same way as we dealt with in the first part of the present chapter.

The motion of the projectile in the x-y plane can thus be thought of as consisting of two independent straight-line motions, which combine to result in the motion we observe.

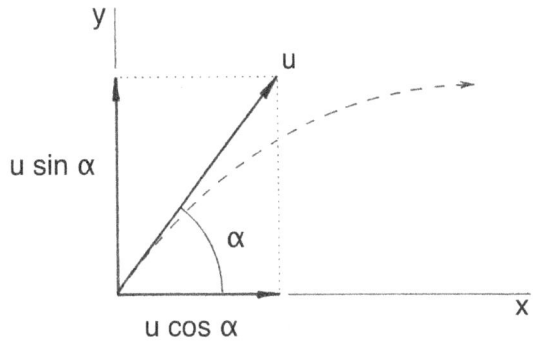

The motion of a projectile that is launched with an initial velocity u at an angle α to the horizontal, can be examined by resolving the initial velocity into vertical and horizontal components:

u sinα and **u cosα** respectively.

The motion in each of these two directions can be analysed separately.

The actual velocity of a projectile at any instant will be the vector sum of its horizontal and vertical velocity components.

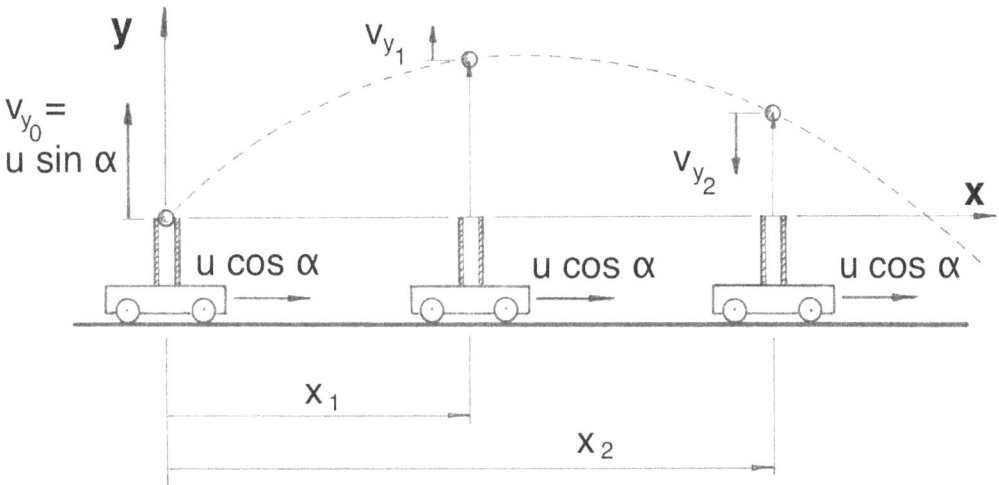

It is useful to imagine the projectile as if it were fired vertically, with initial velocity (u sin α) from a platform that is moving horizontally with a constant velocity of (u cos α).

The x-co-ordinate of the projectile at any instant is equal to the distance travelled by the platform up to that instant. The y-co-ordinate of the projectile may be found by considering the vertical motion of the projectile relative to the platform.

Example

A cricket ball is thrown into the air with an initial velocity of 20 m/s, at an angle of 30° to the horizontal.

Determine its x- and y- co-ordinates relative to the point of launch, 0.5 seconds after being released.

Draw the launching velocity vector, resolved into horizontal and vertical components.

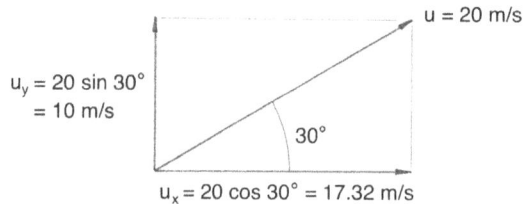

$u_y = 20 \sin 30°$
$= 10$ m/s

$u = 20$ m/s

30°

$u_x = 20 \cos 30° = 17.32$ m/s

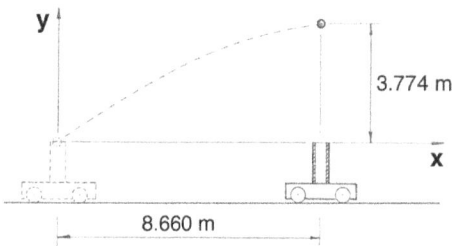

y

3.774 m

x

8.660 m

Think of the ball as if it were launched vertically upwards at 10 m/s from a platform moving to the right with uniform velocity of 17.32 m/s.

Examine the horizontal motion:

The ball moves abreast of the 'platform' with a constant velocity of 17.32 m/s. How far is it from the point of launch after 0.5 seconds?

In the x-direction we have straight line motion without acceleration, so the displacement of the ball is given by: s = ut

∴ s = 17.32 × 0.5 = 8.660 m. This is the x-co-ordinate of the ball at t = 0.5 seconds.

Examine the vertical motion:

The ball moves vertically relative to the 'platform', with an initial velocity of 10 m/s, and is subject to gravitational acceleration.

$s = ut + \frac{1}{2}at^2$ ∴ $s = 10(0.5) + \frac{1}{2}(-9.81)(0.5)^2$ ∴ s = 3.774 m.

This is the y-co-ordinate of the ball at t = 0.5 seconds.

Exercise

For the example above, determine the maximum height and maximum distance achieved by this ball. [5.097 m; 35.30 m]

The platform analogy does not need to be mentioned in the solution of problems. It serves to make us mindful of the fact that the motions in the x- and y- directions need to be treated separately.

General method for approaching all projectile motion exercises

1. Make a sketch of the trajectory, namely the flight path taken by the projectile.

2. Set up the reference frame so that the origin coincides with the point of projection.

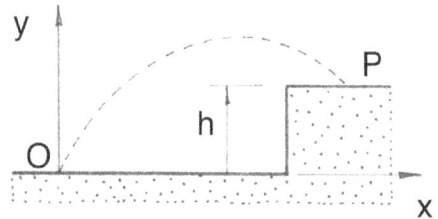

3. Some data pertaining to certain points on the trajectory will be known.
Use this data in a relevant equation for motion in either the x- or the y-direction to determine the values of other variables *pertaining to the same points*.

Exercises on projectile motion: Set 1

Question 1

An arrow is shot with a initial velocity of 40 m/s, at an angle of elevation of 36°. Determine the x- and y- co-ordinates of its displacement from the firing point after 2 seconds.

[x = 64.72 m; y = 27.40 m]

Question 2

A cannon ball is fired with an initial velocity of 200 m/s at an angle of elevation of 40°, over flat ground.

Determine the following, 15 seconds after firing:

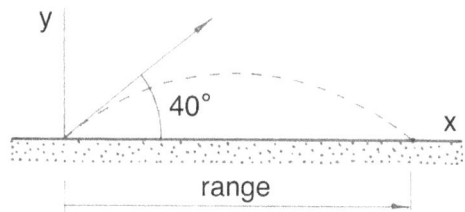

- The height of the ball above the firing point   [824.7 m]

- The vertical component of the ball's velocity   [18.59 m/s downwards]

- The angle at which the ball is moving. (From the resultant of its horizontal and vertical velocities at that instant.)   [6.92 ° downwards from the horizontal]

- Also determine the range of this cannon when fired at this elevation.   [4016 m]

Question 3

The illustration shows one type of ballista used by the Romans, powered by the torsion in twisted rope hanks.

Each separate 'arm' of the 'bow' is inserted into one hank. The bars at the top of the box are then turned until no further twist is possible. The block in the centre of the bow-string is pulled back by a windlass. When the trigger is released, the block accelerates the missile up the channel.

A roughly spherical stone is fired from a ballista on horizontal ground, with a firing velocity of 160 m/s, at an angle of 30° to the horizontal. Determine:

- The time of flight of this stone, from firing to striking the ground [16.31 s],

- The range along the ground [2260 m], and

- The greatest height reached. [326 m]

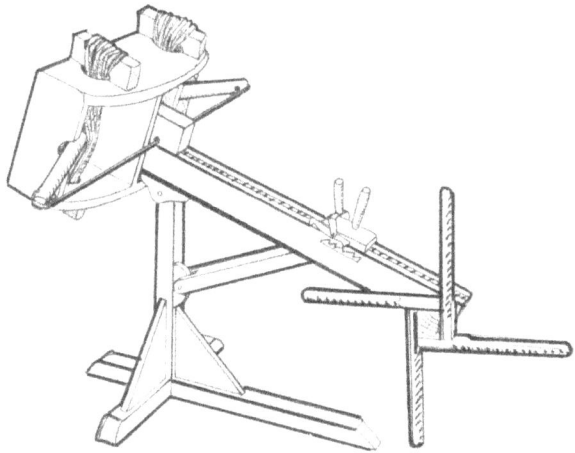

The one shown here was built by Arno de Wet, one of the author's students, for a project whose aim was to launch a tennis ball as far as possible. Height of machine: 1 m. See more on this ballista at the end of this chapter.

Question 4

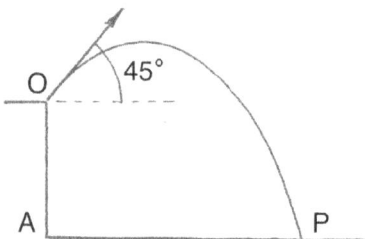

A boy standing at the top of a cliff that is 40 m high tries to throw a stone to land as far as possible from the base of the cliff. He launches the stone with a velocity of 25 m/s, at an angle of 45° upward from the horizontal.

Determine the distance AP. [91.55 m]

Analysing the trajectory of a projectile

If air resistance is ignored, the trajectory is a parabola with a vertical axis of symmetry. It is symmetrical about the highest point reached, H.

At point H, $v_y = 0$, since this is the turning point of the motion in the y-direction.

If fired over flat ground, time taken to reach point H is half the time taken to reach point P.

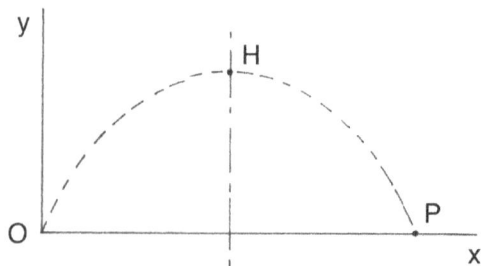

The horizontal range of a projectile over flat ground

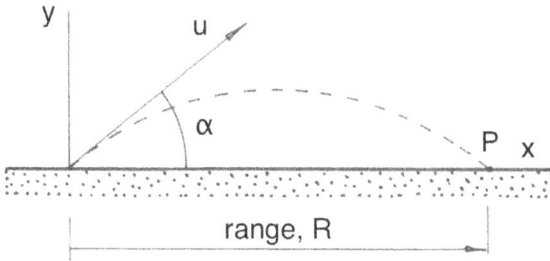

The range, R, of a projectile is the horizontal displacement of the strike point, P.

At point P, the y-displacement of the projectile is zero. Use this fact to determine the time taken to get to point P, and hence the x-displacement of P:

In the y-direction: $s = ut + \frac{1}{2}at^2$ ∴ $0 = u (\sin \alpha)t - 4.905\, t^2$

∴ $t = 0$ (at the point of firing) and $t = u \sin \alpha/4.905$ (at the strike point, P)

Substitute this latter value for t into the equation for the x-displacement of the projectile:

$s = ut + \frac{1}{2}at^2$ but $a = 0$ in the x-direction, ∴ $R = (u \cos \alpha) (u \sin \alpha) /4.905$

Therefore the range, **$R = u^2 (\sin \alpha)(\cos \alpha) /4.905$**...................................(1)

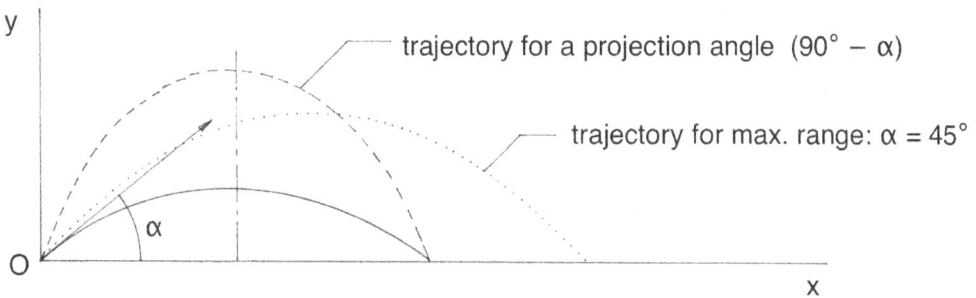

From (1): for any acute angle α: $\cos \alpha = \sin (90° - \alpha)$ so that the expression can also be written as:

$R = u^2 (\sin \alpha) (\sin (90° - \alpha)) /4.905$ ....(2)

Equation (2) indicates that R will have the same value if α is replaced with $(90° - \alpha)$, which means that a projectile with a given firing velocity will have the same horizontal range over flat ground if fired an an angle of elevation of 20° as it will when fired at 70°, and so on.

35

The maximum range over flat ground: substituting the fact that $2(\sin \alpha)(\cos \alpha) = \sin 2\alpha$ into equation (1) above:

$$\therefore R = u^2 (\sin 2\alpha)/9.81 \ldots \ldots \ldots (3)$$

This expression will have a maximum value when $(\sin 2\alpha)$ has a maximum value, namely 1, in which case $2\alpha = 90°$ and thus $\alpha = 45°$.

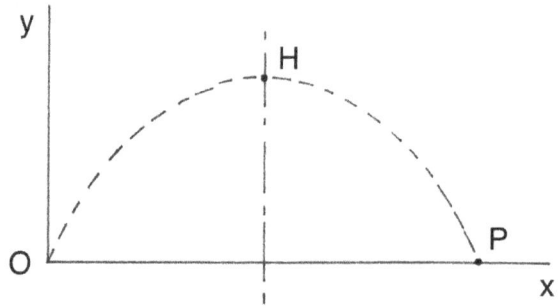

Thus the maximum range over flat ground will be achieved when the angle of elevation of the launch is 45°, and is given by $R = u^2/9.81$

For example, the maximum range over flat ground of a cannon with a muzzle velocity of 180 m/s will be: $R = (180)^2/9.81 = 3303$ m

The equation of the trajectory of a projectile

If air resistance is ignored, as we have seen, the trajectory of a projectile will be a parabola. This parabola can be described by an equation showing y as a function of x.

To derive this equation, we eliminate t from the equations for y as a function of t, and from x as a function of t, as follows:

In the x-direction, from $s = ut + \frac{1}{2}at^2$, $x = u(\cos \alpha)t \therefore t = x \div u (\cos \alpha)\ldots\ldots\ldots$ (1)

In the y-direction, from $s = ut + \frac{1}{2}at^2$, $y = u (\sin \alpha) t - 4.905t^2 \ldots\ldots\ldots$ (2)

Substitute the value of t from (1) into (2):

$y = u(\sin \alpha)(x \div u(\cos \alpha)) - 4.905(x \div u(\cos \alpha))^2$

$$\therefore \mathbf{y = x\ tan\alpha - (4.905/u^2\ cos^2\alpha)x^2} \ldots\ldots\ldots\ldots\ldots\ldots\ldots\ldots (3)$$

alternatively, since $1/\cos^2\alpha = \sec^2\alpha = (1 + \tan^2\alpha)$, equation (3) can also be written as:

$$\mathbf{y = x\ tan\ \alpha - (4.905/u^2)(1 + tan^2\alpha)x^2} \ldots\ldots\ldots\ldots\ldots\ldots (4)$$

Example: An arrow is fired with an initial speed of 100 m/s, at an angle of 42° to the horizontal. How high above the ground will it be, after covering a horizontal distance of 120 m?

Using equation (3) above: $y = 120 \tan 42° - (4.905/ 100^2 \cos^2 42°)120^2 = 95.26$ m

Exercises on projectile motion: Set 2

Question 1

The maximum range of a certain mangonel, a medieval siege machine that hurls rocks, is 600 m on flat ground. Determine its firing velocity. Ignore the extra height above the ground to the point from which the rocks are discharged. [76.02 m/s]

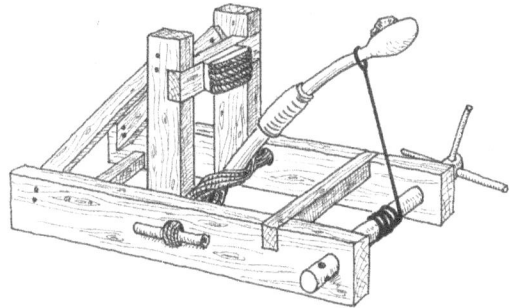

Question 2

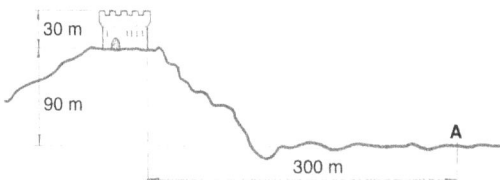

A medieval commander needs to bombard a castle that is on a hill, using the siege machine in question 1 above. The closest his men can set up the machine is at position A.

Assuming his estimates of the dimensions shown are correct, if the rocks are launched at an elevation of 35°, are they likely to hit the walls? [Yes]

Question 3

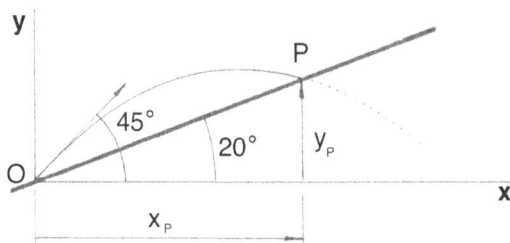

A cannon with a muzzle velocity of 150 m/s is required to fire up a slope of 20°.

If it fires at an elevation of 45° to the horizontal, what is its range up the slope (distance OP)? [1552 m]

Question 4

A flow of gravel is discharged from a chute at an angle of 20° below the horizontal, at a speed of 5 m/s.

If the gravel is to land on a conveyor belt situated 2 m below the mouth of the chute, what is the average distance 'd' from the mouth of the chute that the gravel will land on the belt? [2.291 m]

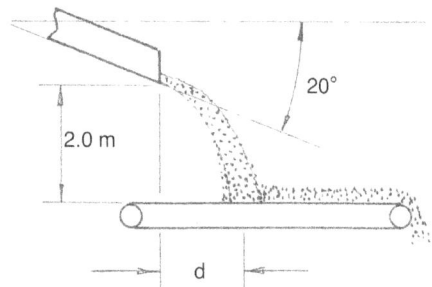

A practical way to determine the launch velocity of a high-speed projectile

There is an relatively simple way to determine the initial velocity of an arrow, pellet, bullet, or *small* cannon ball. Larger missiles are excluded from this analysis on account of the fact that the damage they cause precludes accurate measurement of their strike point on a target.

The method is based on the fact that we can consider the horizontal and vertical motions of a projectile as separate.

The firing device (crossbow, rifle, airgun or small cannon) is set up to be firmly fixed and the axis of the barrel is lined up exactly horizontally, pointing at a target which is at the same elevation, and a known distance away. The target should have an absorbent backing, to retain the projectile, enable the strike point to be established, and reduce the likelihood of danger to experimenters.

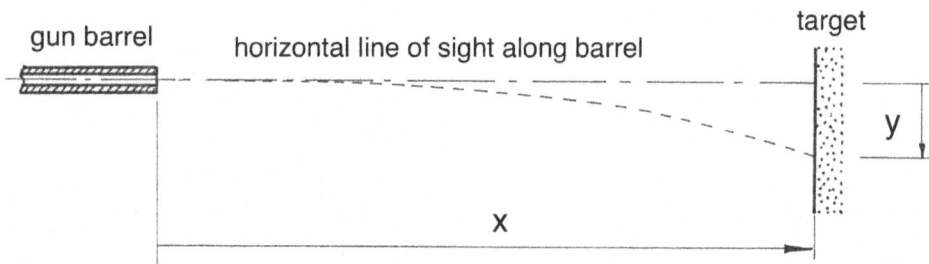

When the device is fired, the projectile will hit the target at some point directly below the bullseye. The dimension 'y' between the centre of the bullseye and the actual point of strike is measured. The greater the launching velocity, the smaller this dimension will be.

The vertical component of the projection velocity is zero, so that, in the y-direction, the projectile behaves as if it were dropped from rest. In the x-direction, it has a firing velocity of u, which remains constant throughout the flight (disregarding air resistance).

Suppose that, for a crossbow, tested in this way, the distance to the target is 10 m, and the bolt strikes the target 65 mm below the centre of the bullseye.

In the vertical direction, from $s = ut + \frac{1}{2}at^2$, since $u = 0$, we have $-0.065 = -4.905t^2$

$\therefore t = 0.115116$ seconds

In the horizontal direction, using the same equation, namely $s = ut + \frac{1}{2}at^2$, and noting that there is no acceleration in this direction:

$10 = u(0.115116) \therefore u = 86.87$ m/s

This method, though valid for all projectiles, would present a challenge if used for arrows shot from a bow, or for balls thrown by hand, as it would be difficult to ensure that such projectiles are launched exactly horizontally.

The amount of drop, namely dimension 'y', for a given range, x, allows us to set the rangefinders on rifles. The slower the projectile, the more one has to compensate for the anticipated drop. This effect is particularly noticeable in archery.

Exercise

A rifle is set up in a vice so that its barrel is lined up precisely horizontally, aiming at a target that is 50m away. A shot is fired and hits the target 32 mm below the bullseye.

Estimate the firing velocity.   [619.0 m/s] (Why is this only an estimate?)

Determine the angle of elevation required for this rifle to hit a target at the same elevation as the barrel, at a range of 200 m. [0.000733°]

If the rifle barrel is 600 mm long, by how much should the front end of the barrel be set higher than the breech end to correct the trajectory for this range? [0.0077 mm]

The effect of air resistance on the trajectory of a golf ball

In all the analysis presented above, air resistance has been ignored. However, we know that air resistance is significant. The question then arises: how significant? Some indication of the extent to which air resistance affects our calculations can be inferred from the following experiment, in which the author and his students used a purpose-built apparatus to plot the actual trajectory of a golf ball over a maximum range of 3 m.

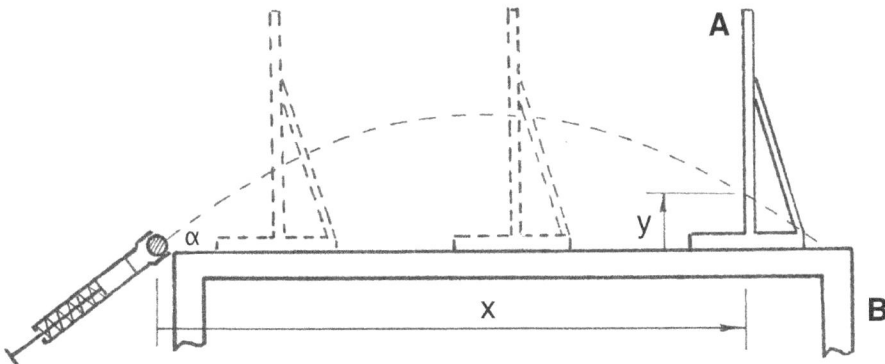

The apparatus consisted of a spring-gun, triggered by switching off the current to an electromagnet that held the plunger in place against the compressed spring. The plunger struck a golf ball out of a machined cup, at a measurable angle α, with controlled velocity that gave it a maximum range of approximately 3 m.

The spring gun was mounted at one end of a table (B) with a horizontal surface. Vertical panel A could be placed in various positions with known x-co-ordinates along the table.

The angle α was set and the ball was fired with a velocity that would not take it beyond the end of table B. The vertical panel A was covered with paper, over which a sheet of carbon paper was taped. Each strike made a mark on the paper, so that the vertical position of the ball could be co-ordinated with its horizontal distance from the launching point. Panel A was moved to different positions for successive shots, enabling the trajectory to be plotted.

Over many experiments carried out on many different occasions, by different groups, it was found that the trajectory for a launching angle of 45° fell short of a symmetrical parabolic trajectory by approximately 8% in the horizontal range. The symmetrical parabola was constructed using the co-ordinates of three plot points from the first half of the actual trajectory.

Naturally, the ideal parabola in such cases could not be determined from the observations, since the vertical motion would also have been affected by air resistance. To describe the ideal trajectory, it would have been necessary to determine the launching velocity accurately, and then to use that value in an equation for the trajectory. All we could conclude from this experiment was that the trajectory was *not* symmetrical about the highest point reached. We could measure by how much the horizontal range had been reduced, assuming that the second half of the trajectory ought to have been a mirror image of the first half, which was estimated to be close to parabolic.

The observed difference between the actual trajectory and a parabola could only be ascribed to the effect of air resistance.

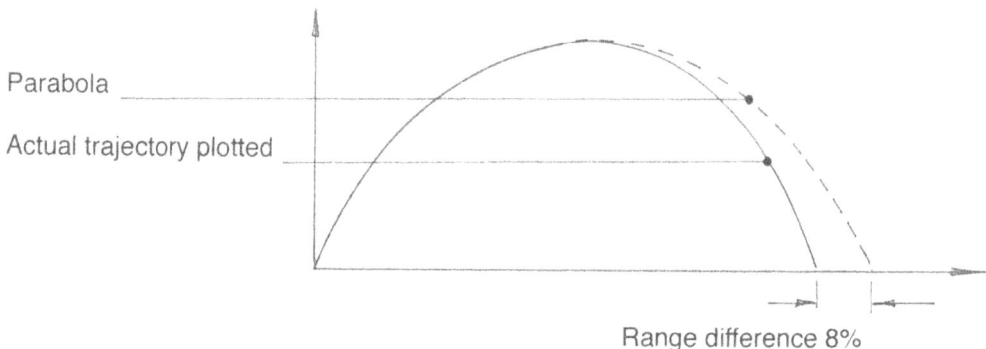

Parabola

Actual trajectory plotted

Range difference 8%

In this experiment the magnitude of the launching velocity was small.

The reader should consider whether higher velocities would result in greater or lesser range differences.

The effect of air resistance on the flight of a ball is not always undesirable. For example, the drop at the end of the flight greatly assists golfers, since, if a ball were to return to ground at the same angle from which it was struck, it would bounce with a large horizontal component of velocity, resulting in its placement being extremely difficult to control.

Exercises on projectile motion: set 3

Question 1

A cannon ball with muzzle velocity of 200 m/s is fired at a target 2200 m away, 300m above the firing point. Determine the angle of elevation that will enable the ball to hit the target.   [72.86° or 24.90°. *Note: these two values can not be expected to add up to 90°, which would be the case if the projectile was fired over flat ground.*]

Question 2

A catapult shoots a stone with an initial velocity of 25 m/s. Determine the range of values of the projection angle that will result in the stone landing on the flat roof of an 8 m high building, 30 m from the point of projection, and 10 m deep.

[Between 30.19° and 32.79° and also, between 68.52° and 74.74°]

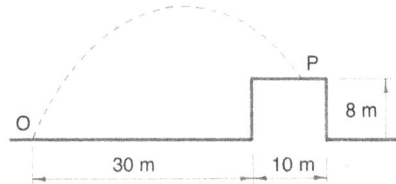

Question 3

A long jumper's centre of gravity is 1.2 m above the ground when running, and 0.4 m above the ground when landing. If he takes off with a forward speed of 10 m/s and jumps a distance of 7.00 m, determine:

- His initial velocity component in the vertical direction   [2.29 m/s],

- His angle of projection   [12.90°],

- The maximum height above ground reached by his centre of gravity during the jump   [1.467 m], and

- The distance he could jump if he could manage to launch himself with a vertical velocity component of 3 m/s while maintaining a forward velocity of 10 m/s on take-off.   [8.12 m]

Question 4

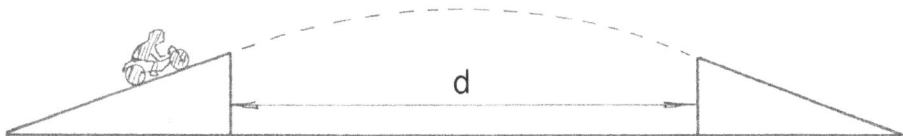

Motorcycle ramping distance: Determine the ramping distances 'd' for each of the following take-off speeds, assuming the bike lands at the same elevation as the one

41

from which it was launched, and that the ramp is inclined at 20° to the horizontal:

Determine also, in each case, the maximum height above the top of the take-off ramp reached by the lower surface of the bike wheels. Some of the answers are given in the table below. Confirm these and complete the table.

Speed [km/h]	36	54	72	90	108	126
Speed [m/s]	10			25		
Ramping distance 'd' [m]	6.552			40.95		
Max height reached [m]	0.596			3.726		

Additional exercises: vertical motion and projectile motion

The following exercises are suitable for discussion in a small group. Solutions deliberately not supplied.

Question 1

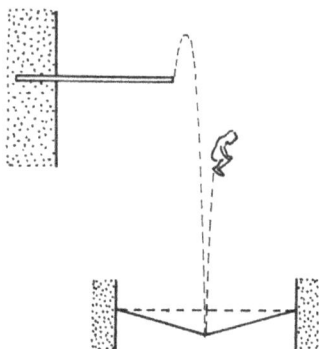

Sketch a v - t graph for the motion of a trampolinist who bounces upwards from a diving board to a height of 2m above the board, and then falls onto a trampoline 10 m below the diving board, after which he bounces up to a height of 7 m above the level of the trampoline.

The trampoline skin stretches a distance of 1 m below its normal level at its maximum stretch in his downward movement. Indicate on the graph all displacements incurred at the various stages of the movement. Assume all accelerations are uniform.

Question 2: designing an experiment

Design an experiment to determine the muzzle velocity of a pellet from an airgun. Describe how you would:

• Secure the airgun in a fixed firing position,

• Ensure that the barrel is lined up perfectly with the bullseye on exactly the same horizontal level,

• Choose a firing distance to the target,

• Ensure the safety of the experimenters,

- Measure the downward deviation of the pellets on the target,

- Choose a number of pellets to fire to take a reasonable average reading, and

- Compensate for heat build-up in the airgun affecting the outcome.

Question 3: A practical experiment

Estimate the launching velocities of the following types of projectiles by timing how long it takes for them to return to the level from which they were projected.

In each case the projectile should be launched as near to vertically as possible. A small ball thrown, a soccer ball kicked as high as possible, a tennis ball struck with a racquet, a small stone fired from a catapult, an arrow fired from a relatively weak bow, and one fired from a stronger bow.

Take proper safety precautions.

Notes on how the ballista illustrated previously works, with questions

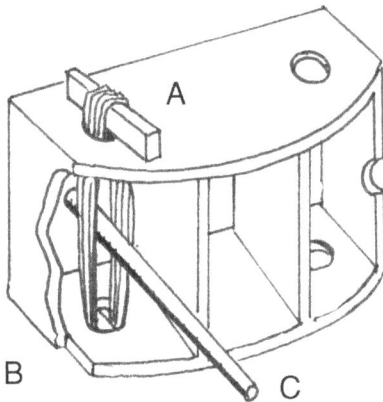

1. Place bar A and its counterpart B over the outside of the holes in the box. Bar B is below the bottom hole in the box.

2. Thread a length of rope over A and B, through the holes, to form a hank. Tie firmly.

3. Insert the 'bow' arm C through the hank.

4. Turn bars A and B clockwise as seen from above, until they can't be turned any further. The twisted hank now presses arm C against the side of the box.

5. Bars A and B are kept in place by friction with the outer surface of the box.

6. It will take a significant force to pull arm C in the direction 'x', due to the torsion in the twisted hank of rope.

7. Arm C is now connected to its counterpart on the other side of the box by a strong 'bow-string' cord, that passes through a light block riding in the channel.

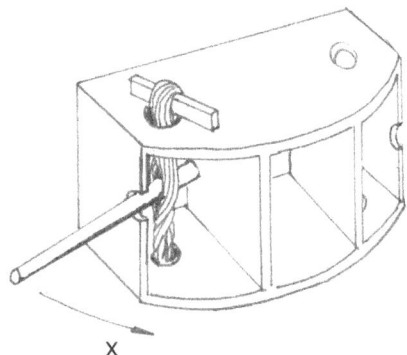

8. The block is pulled down the channel by means of a rope and windlass.

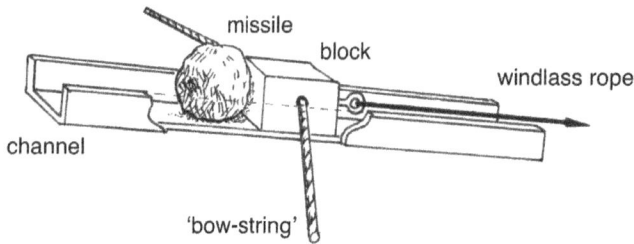

9. The missile is placed in front of the block. The block is released from the windlass rope by means of a trigger. The block accelerates the missile up the channel.

Design questions for discussion, relating to this ballista

a. What type of rope would work best for the twisted hanks: nylon or natural fibre? Why?

b. What characteristics would you look for in choosing material for the 'bow' arms?

c. How would you absorb the shock of the 'bow' arms hitting the sides of the box at the end of a shot?

d. Where does this mechanism appear to be inefficient, in that energy is lost unnecessarily?

e. What could be done about this?

f. What should be the characteristics of the material used for the 'bowstring'?

g. Should the block preferably be light or heavy?

h. Suggest a means of ensuring that the block cannot be released inadvertently while winding the windlass.

i. Suggest a trigger mechanism for releasing the block from the windlass rope.

j. If the missile were a tennis ball, would it roll on its way up the channel? Why, or why not?

k. If it did roll, what would be the effect of imparting such spin to the projectile?

l. If the missile were a rounded stone, what implications does that have for the material of which the channel is made?

m. How would you ensure that the channel does not wear excessively? What are the respective pros and cons of (1) making it from harder material, or (2) lining it with a removable, disposable lining?

Chapter 12

Rotational motion

Definition of a radian, angular displacement, angular velocity and acceleration.
Relations between angular quantities and their linear counterparts.
The velocity-time graph for angular motion
Velocity ratios and acceleration ratios of gears and belt drives.

The earliest known depiction of a wheel used in transport comes from a Sumerian mosaic of 4600 years ago, showing chariots and wagons in a royal procession.

Wheels, and their derivatives: shafts, gears, pulleys, flywheels, armatures, turbines, propellers, impellers, all rotate. It is therefore essential for an engineer to understand the basic relations that apply to rotary motion.

It is possible that wheels were used for wagons and making pottery even before the Sumerian civilisation. Waterwheels and windmills have been in use for thousands of years. Naturally, as technology has developed, rotating items have found many more applications, and today are seldom made of wood, but are made of metal or other advanced materials, and are often required to rotate at high speed, and sustain high levels of stress.

In order to understand rotary motion, we must be familiar with the unit of angular measurement known as the radian.

The definition of the radian as a measurement of angle

The most familiar and universally used unit of angular measurement is the degree. One full circle is divided into 360 degrees. This is not coincidental: it originates in the fact that 360 is divisible by 3, by 4, by 5 and by 6. (3 × 4 × 5 × 6 = 360). This number is also divisible by 8, 9, 10 and 12, making it the most suitable number for the various purposes that people have found for dividing up a circle.

Even though it is never likely to wane in popularity, the degree is an arbitrarily chosen measure of angle. Equally arbitrarily, instead of dividing a right angle into 90 divisions, one could have made it 100 divisions. In fact, some people have done this, and called their divisions 'gradians'. Though your calculator will accommodate this option, engineers seldom need it, as the use of such a measure seems to be confined to land surveyors in Europe, particularly in France, where it was developed as part of the metric system.

For many purposes in engineering, we need, not an arbitrary, but a *naturally occurring* measure of angle. This measure is called a radian.

A radian is defined as that angle for which the arc length, s, is equal to the length of the radius, r. Consider an angle θ that subtends an arc of length s, in a circle of radius r.

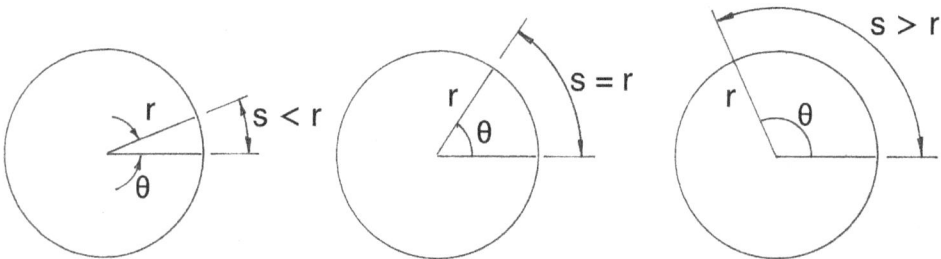

Here θ is a small angle   Here θ = 1 radian   Here θ >1 radian

To get a perspective on the size of a radian, in degrees, compare the arc length, s, with the radius, r, when θ = 60°, forming an equilateral triangle.

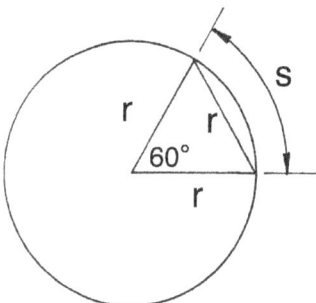

In this case the arc length is clearly slightly longer than the radius.

If the arc length were to be *equal* to the radius, the angle θ would have to be slightly smaller than 60°. In fact, as will be seen below, one radian is approximately 57.3 °.

To see how this is derived, we need to establish a general relation between an angle in radians and the arc length that it subtends.

In order to develop an expression relating arc length, angle subtended, and radius, consider the following three diagrams:

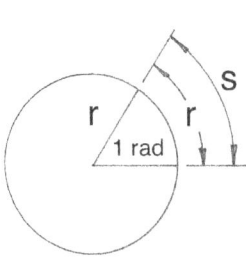 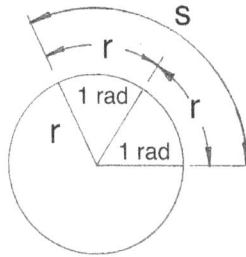 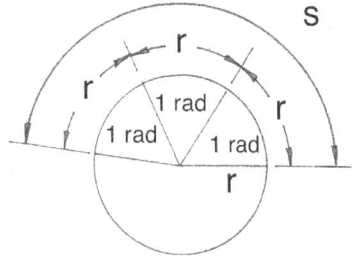

If θ = 1 rad, s = r　　　　If θ= 2 rad, s = 2r　　　　If θ = 3 rad, s = 3r

So, arc length s equals the radius times the number of radians:　　**s = r θ**

The number of radians in one full revolution

The length of the circumference of a circle is 2 π r. The arc length, s, of one full circle, containing θ radians, is given by s = r θ, so, for one complete circle:

∴ 2 π r = r θ　∴　θ = 2 π　 Hence, there are 2π radians in one full circle.

And, since one full revolution contains 360° :

360° = 2 π radians　∴ 1 rad = 360° ÷ (2 π) = 57.296°

Example

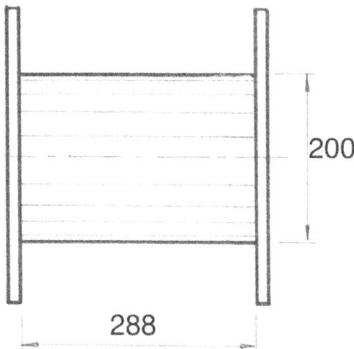

A reel onto which rope must be wound has diameter 200 mm. The width between the flanges of the reel is 288 mm. How much rope of diameter 8 mm can be wound onto the reel in the first layer of windings?

The effective radius of each winding is the radius of the reel plus half the thickness of the rope, namely 104 mm. Each winding takes up an arc length of rope equal to:

s = rθ = 104(2π) = 653.5 mm.

The number of windings in the first layer is 288 ÷ 8 = 36, so the total length of rope in the first layer will be 653.5 × 36 = 23524 mm = 23.52 m.

Example

One way to measure distance along uneven ground is to make use of a surveyor's wheel. This is a rod with a small wheel attached, that can be pushed or pulled by a person walking. A mechanism records the number of revolutions turned by the wheel. If the wheel has diameter 220 mm and registers 569 revolutions, how much ground has been covered? Assume the wheel does not slip on the ground.

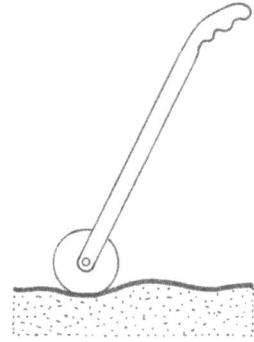

$s = r\theta = (0.110)(569 \times 2\pi) = 393.3$ m

Exercises on radian measure, set 1

1. If a wheel such as that in the example above must record exactly one metre for each revolution turned, what should be the effective radius of the wheel? [159.2 mm]

2. A circular water tank has a circumference of 40 m. Two points on the tank wall are 3 m apart. What angle will this arc subtend at the centre of the circle? [0.4712 rad]

3. If the average radius of the Earth is 6371 km, how many km apart are two points on the equator that have a 1° difference in longitude? [111.2 km]

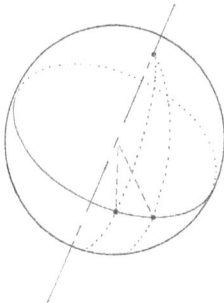

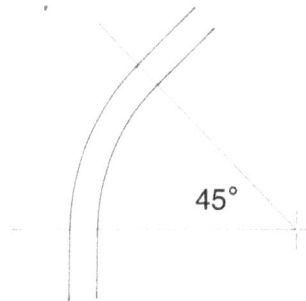

4. A circular curve in a railway on horizontal ground has an average radius of 200 m, and the rails are 1.2 m apart. This curve changes the direction of the track by 45°. By how much will the outer rail in the curve be longer than the inner rail? [0.9425 m]

The definition of angular displacement, θ (theta)

Recall that the *linear* displacement of an object is the distance from a specified reference point to where an object may be found, in a specified direction. Angular displacement is similarly defined.

The angular displacement of a rotating object is the number of radians by which its angular orientation differs from a specified initial orientation.

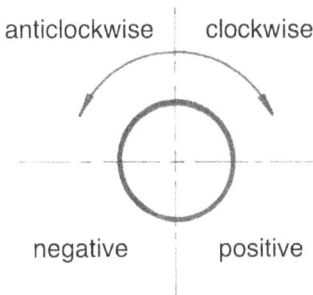

anticlockwise | clockwise

negative | positive

As with linear motion, when dealing with angular motion it is essential to specify whether the change in angle has occurred in one direction, or in the opposite direction.

It is an arbitrary, but useful, convention to regard clockwise as the positive direction. This convention echoes the system used in navigation, where bearings are measured clockwise from north. It really makes no difference if you use this convention, or the mathematical one of measuring angles anticlockwise from east, as long as you are consistent.

With the convention used in this book, if a wheel, starting in a particular position, turns clockwise through 4 radians, its angular displacement from the starting position would be stated as + 4 radians.

The definition of angular velocity, ω (omega)

The rotational *speed* of a rotating object is the rate at which it turns through angle. In common usage, the rotational speeds of engines, lathes, turbines, circular saws and grinders are usually expressed in revolutions per minute, sometimes designated as r.p.m. and sometimes as r/min. However, to be compatible in all calculations that relate angular velocity to other quantities in mechanics, angular velocity should be specified in the basic S.I. unit of radians per second [rad/s].

The definition of angular *velocity* (as distinguished from angular or rotational *speed*) is the rate of change of angular *displacement*. There is a direct parallel here with *linear* velocity, which is the rate of change of *linear* displacement.

If a wheel is rotating at a constant speed in a clockwise direction, at a rate of 5 radians every second, we would say its angular velocity is + 5 rad/s.

Converting radians per second to revs per minute, and vice versa

Example: A shaft turns at 72 r/min. Express this velocity in rad/s.

72 rev/min = 72 rev/min × 2π rad/rev × 1/60 min/s = 7.54 rad/s

Hence, to convert r/min to rad/s, multiply by 2π/60

Example: Express an angular velocity of 518 rad/s in r/min:

518 rad/s = 518 rad/s × 60/1 s/min × 1/2π rev/rad = 4947 rev/min

∴ To convert rad/s to r/min, multiply by 60/2π

The relation between the linear speed of a point on a rotating object and the angular velocity of the rotation.

If an object rotating at *constant* angular velocity ω turns through angle θ radians in time t seconds, we can equate $\omega = \theta/t$ rad/s. If a point on that object, situated at radius r from the centre of rotation, is covering s linear metres in t seconds, its linear speed, v = s/t metres per second. Now, the arc length, s, covered in a given time, t, is related to the amount of angle turned through in that time by:

$$s = r\theta \quad \therefore v = (r\theta)/t \;=\; r\,(\theta/t) \quad \text{therefore} \quad \mathbf{v = r\omega}$$

Example

A long-playing vinyl record of diameter 300 mm turns at 33⅓ r.p.m. What is the linear speed of a point on the rim of the record?

33⅓ r.p.m. = 33⅓ rev/min × 2π/60 = 3.4907 rad/s

$v = r\omega \quad \therefore v = 0.15(3.4907) = 0.5236$ m/s

Example

A mine 'cage' is an elevator used to raise and lower men and materials in a mine. The cage is attached to a steel wire rope that is wound onto a winding drum. If a given winding drum has diameter 4 m, and the velocity of the cage is not to exceed 36 km/h, what should be the maximum angular velocity of the winding drum, expressed in r/min?

Always work in the basic units of the SI system, and convert the answers into other units if required.
36 km/h = 10 m/s

$v = r\omega \quad \therefore \omega = v/r \quad \therefore \omega = 10 \div 2 = 5$ rad/s

$\therefore \omega = 5 \times (60/2\pi)$ r/min = 47.75 r/min

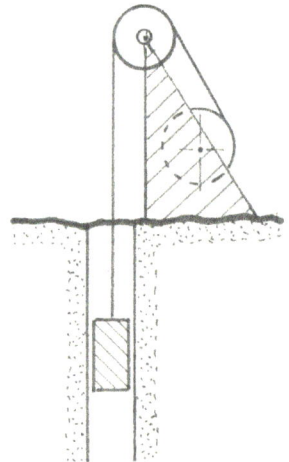

Exercises, set 2

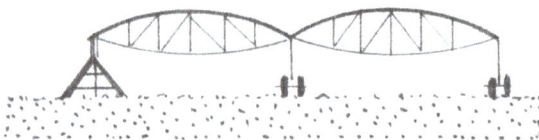

1. The radial arm of a circular irrigation system on a farm rotates slowly, as it sprays. If it does one full turn at constant speed in 2 hours, and the arm radius is 50 m, what is the linear speed of a point on the outer end of the arm?   [43.6 mm/s]

2. Our Earth is on average 93 million miles from the sun. It does one full orbit in one year. With what speed is the planet travelling along the path of its orbit? Assume a circular orbit.   [approx. 107 300 km/h]

3. What is the angular velocity of our planet in its orbit? [0.1991×10^{-6} rad/s]

4. The Earth spins one full revolution on its axis in 24 hours. What is the linear speed of a point on the equator, relative to the Earth's axis? Earth's average radius = 6371 km. [1676 km/h]

5. A pulley wheel in a machine, of diameter 160 mm, revolves at 200 r/min without stopping for a period of exactly 60 days. How far has a point on the edge of the wheel moved in that time?   [8 686 km]

6. For machining a given material on a lathe, the recommended cutting speed is 2.7 m/s. If you are machining at a cut radius of 38 mm, at what angular velocity should the chuck be turning?   [678.5 r/min]

7. A rope is wound onto a winch drum of diameter 250 mm. The drum is turned by a crank handle of radius 360 mm. How many revolutions must you turn the crank to wind up 12.5 m of rope?   [15.92]

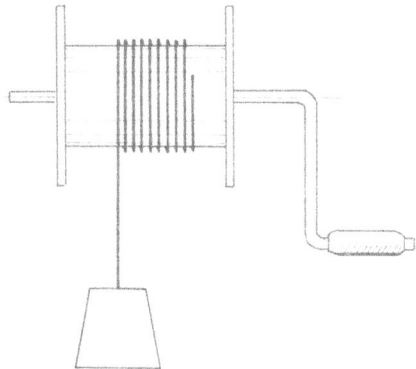

The definition of angular acceleration, α (alpha)

When you set off on a bicycle, from a standing start, and gradually speed up, the bicycle moves forward with linear acceleration. At the same time, the wheels are also speeding up in their angular motion. They are undergoing angular acceleration, which is the rotational counterpart of linear acceleration.

Angular acceleration, α, is defined as the *rate of change of angular velocity*. Any rotating object that is speeding up is experiencing angular acceleration. If it is slowing down, it is experiencing angular *deceleration*.

By the above definition, if the rotating object increases its angular velocity by an amount $\Delta\omega$ in a time Δt, then the angular acceleration $\alpha = \Delta\omega/\Delta t$

The relation between the angular acceleration of a rotating object and the 'linear' tangential acceleration of a point at some radius, r, on the object.

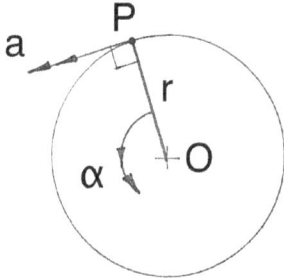

If line OP is moving with angular velocity ω, then point P is moving along the circumference of the circle with 'linear' speed v, where $v = r\omega$.

If point P is experiencing *linear* acceleration along the circumference of the circle, then line OP is experiencing *angular* acceleration.

The linear acceleration of P and the angular acceleration of line OP are directly related.

As on the above diagram, from now on, we will use double-headed arrows for accelerations, and single-headed arrows for velocities. This will apply to both linear and rotational motion.

The 'linear' acceleration of point P can be expressed as the rate of change of speed, v, with time, t, as follows: $a = \Delta v/\Delta t$ (the change in velocity divided by the change in time)

But $v = r\omega$, so $\Delta v = r\,\Delta\omega$ (noting that r is a constant) and thus we have

$a = (r\,\Delta\omega)/\Delta t = r\,(\Delta\omega/\Delta t)$ therefore: **$a = r\alpha$**

Example

A cord is wound around a drum (ϕ 300 mm) with a horizontal axis of rotation. A mass-piece is attached to the free end of the rope. This mass-piece is allowed to descend, from rest, causing the drum to turn. It takes 5 seconds to unwind 10 m of the cord. Determine the acceleration of the mass-piece and the angular acceleration of the drum.

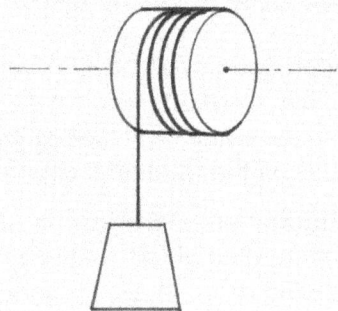

Firstly, the linear motion of the mass-piece, accelerating from rest, covering a distance of 10 m in 5 seconds:

From **$s = ut + \tfrac{1}{2}at^2$**, we obtain: $a = 0.8000$ m/s².

From **$a = r\alpha$**, we have $\alpha = 0.8/0.15 = 5.333$ rad/s²

52

Example:

An antique treadle-powered grinding wheel of diameter 800 mm starts at rest, and is accelerated by the operator until a point on the rim has reached the operating speed of 5 m/s. This process takes 10 seconds. What is the angular acceleration of the wheel?

The angular velocity when the wheel has reached operating speed:

$v = r\omega$ ∴ $\omega = v/r = 5/0.4 = 12.5$ rad/s

If the acceleration has been uniform, then
$\alpha = \Delta\omega/\Delta t = 12.5$ rad/s ÷ 10 s = 1.25 rad/s^2

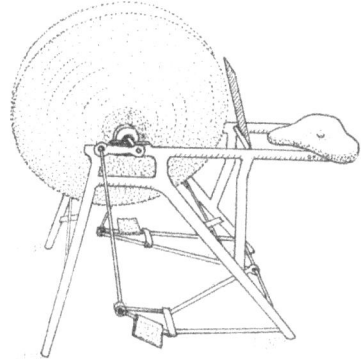

Relations between the angular quantities and their linear counterparts

The angular displacement velocity, and acceleration of a rotating object are related by equations which have *exactly the same form* as do the equations governing the relations between the equivalent quantities in linear motion.

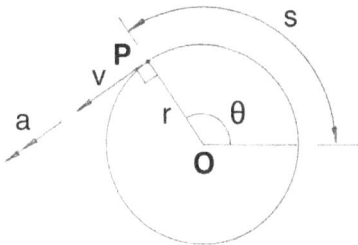

Consider a point P moving in a circle of radius r. The linear displacement, velocity and acceleration of point P can be thought of as applying to a straight line path which has been rolled up onto the circumference of this circle.

For the 'linear' motion of point P on this rolled up path, the three equations of straight line motion with uniform acceleration apply, namely:

v = u + at .... (1)

s = ut + ½at² .... (2)

v² = u² + 2as .... (3)

Each linear quantity in these equations is related to an angular quantity pertaining to the line OP, by the relations that we have seen above, namely:

s = rθ .... (4)

v = rω .... (5)

a = rα .... (6)

Substituting the values from equations (5) and (6) into equation (1):

$v = u + at$ ∴ $r\omega_2 = r\omega_1 + (r\alpha)t$

Since r is a common factor to all the terms in this equation: $\omega_2 = \omega_1 + \alpha t$ which is an equation of exactly the same form as its linear counterpart, equation (1) above.

By a similar process, we can show that equations (2) and (3) have directly similar-looking counterparts in angular motion, as illustrated in the following table:

Linear motion equation	linking	Angular motion counterpart equation
$v = u + at$	initial velocity final velocity acceleration time	$\omega_2 = \omega_1 + \alpha t$
$s = ut + \frac{1}{2}at^2$	displacement initial velocity acceleration time	$\theta = \omega_1 t + \frac{1}{2}\alpha t^2$
$v^2 = u^2 + 2as$	initial velocity final velocity acceleration displacement	$\omega_2^2 = \omega_1^2 + 2\alpha\theta$

It is not necessary to take trouble to memorise the three equations deduced above. They may be written out by simply replacing the linear quantities in the well-known equations (1), (2) and (3) with their corresponding angular quantities.

The similarity between the linear and angular equations allows the use of similar methods for solving problems in angular motion, to those used for linear motion.

The velocity-time graph for angular motion

The linear velocity-time graph and the angular velocity-time graph are similar:

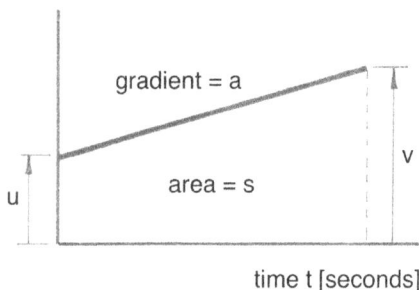

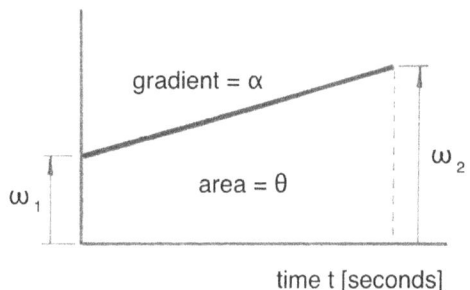

As we did when solving problems in linear motion, when dealing with a problem to do with angular motion, we can:

- Work from the dimensions of a velocity-time graph, and/or
- Use the equations of angular motion with appropriate initial values once for each unique period of motion.

Example

A wheel of radius 1.5 m is initially at rest. It is rotated with uniform angular acceleration of 2 rad/s^2 until it reaches an angular velocity of 12 rad/s, and then continues at constant velocity for a further 8 seconds, after which a brake is applied and it slows down with uniform deceleration until it comes to rest in a further 12 seconds.

a. How long does it take to reach the angular velocity of 12 rad/s?

b. How many revolutions does it turn through in the entire motion?

First, sketch the velocity-time graph of this motion:

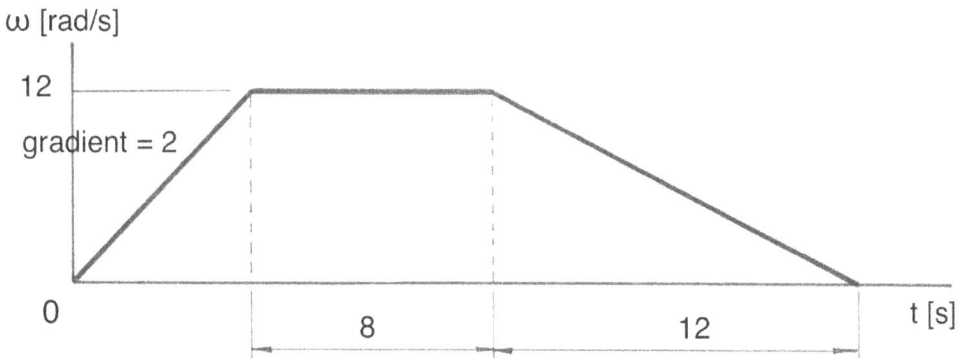

There are three separate periods of motion here, each with its own value of uniform angular acceleration.

For the period of acceleration: $\omega_1 = 0$; $\omega_2 = 12$ rad/s; $\alpha = 2$ rad/s^2
From $\omega_2 = \omega_1 + \alpha t$: $12 = 0 + 2t$ $\therefore t = 6$ seconds
Angular displacement, in radians, during this period:
From $\theta = \omega_1 t + \frac{1}{2}\alpha t^2$: $\theta = 0 + \frac{1}{2}(2)\, 6^2 = 36$ radians

For the period of constant velocity: $\omega_1 = 12$ rad/s; $\omega_2 = 12$ rad/s; $t = 8$ s; $\alpha = 0$ rad/s^2
From $\theta = \omega_1 t + \frac{1}{2}\alpha t^2$: $\theta = 12\,(8) + 0 = 96$ radians

For the period of deceleration: $\omega_1 = 12$ rad/s $\omega_2 = 0$ rad/s and $t = 12$ seconds
From $\omega_2 = \omega_1 + \alpha t$: $0 = 12 + \alpha(12)$ $\therefore \alpha = -1$ rad/s^2
And $\theta = \omega_1 t + \frac{1}{2}\alpha t^2$: $\theta = 12(12) + \frac{1}{2}(-1)12^2 = 72$ radians

The total angle through which the wheel rotated, including all three periods of the motion, is 36 + 96 + 72 = 204 radians. This represents 204 ÷ 2π = 32.47 revolutions.

The above answers can be checked by examining the corresponding features on the velocity-time graph. Divide the graph into sections representing the three periods of motion:

The period of acceleration:

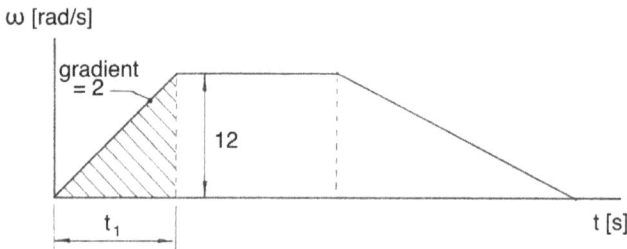

Since gradient = vertical ÷ horizontal, 2 = 12 ÷ t_1

∴ t_1 = 6 seconds.

The angular displacement is equal to the area under the graph, namely ½(6)(12)

= 36 radians

The period of constant velocity

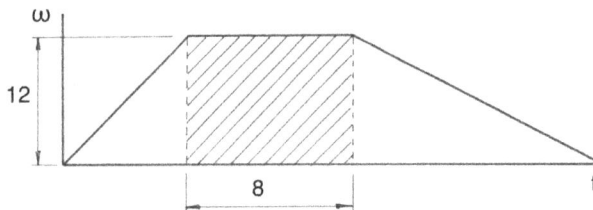

Angular displacement incurred during this period = area under the graph

= 12 × 8 = 96 radians

The period of deceleration

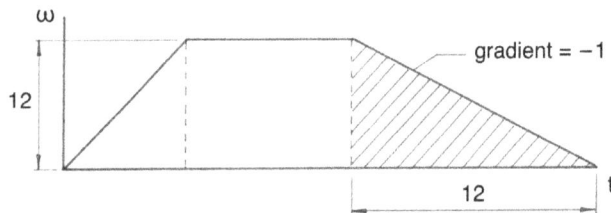

The gradient represents the value of the angular acceleration.

Gradient

= vertical ÷ horizontal

= (– 12) ÷ (12) = –1 rad/s²

Angular displacement during this period = area under the graph = ½(12)(12) = 72 rad

So, working with the geometry of the ω – t graph produces the same result as working with the equations governing rotational motion with uniform acceleration.

Exercises, set 3: angular motion with periods of acceleration

Question 1   A wheel of radius 2 m, mounted on an axle and free to turn, starts from rest and is given an angular acceleration of 4 rad/s^2. After five seconds, what is:

- The angular velocity of the wheel [20 rad/s], and
- The 'linear' speed of a point on the rim of the wheel?   [40 m/s]

Question 2

A winding drum for raising and lowering the cage in a mineshaft has diameter 4 m. It begins from rest and turns with constant angular acceleration of 2 rad/s^2 for 15 seconds, lowering its wire rope down the shaft.

It is then braked with uniform angular deceleration of 3 rad/s^2 until it stops. How much rope has been let out, assuming all the rope was wound onto the drum at a diameter of 4 m, to begin with?   [750 m]

Question 3

A wheel of radius 800 mm is rotating with uniform angular velocity, such that the linear velocity of a point on the rim of the wheel is 1.2 m/s. At a given instant a brake is applied, and the wheel is subjected to a uniform deceleration of 0.2 rad/s^2. Determine:

- How long it takes for a point on the rim to reach a speed of 0.4 m/s, [5 sec] and
- The number of revolutions it turns while slowing down from the initial speed to 0.4 m/s. [0.796 revs]

Question 4

A machine used to subject astronauts to high 'g' forces works by spinning them around at the end of a horizontally rotating arm.

If the machine takes 18 seconds from rest (with uniform acceleration) to perform its first revolution:

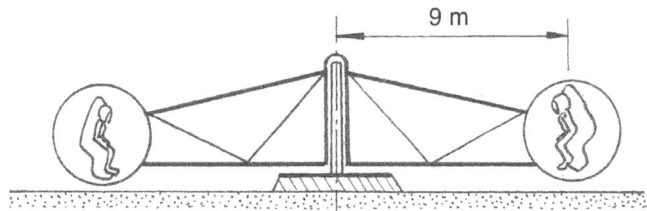

- What is the linear acceleration of the astronaut's seat along the circumference of the circle of movement? [0.3491 m/s^2]
- What is the angular acceleration of the machine arm? [0.03879 rad/s^2]
- If the machine continues to accelerate at this rate, how many revolutions after starting will it take for the end of the rotating arm to reach a linear speed of 25 m/s? [15.83 revs]

Velocity ratios and acceleration ratios of gear and belt drives

Many engineering applications require the transfer of rotary motion from a motor shaft to that of a driven machine, either by gearing, or by a system of pulleys and belts, or by a chain and sprockets.

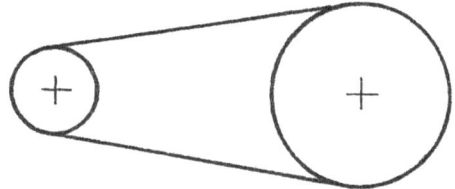

In this section, we present the basic reasoning that allows us to deduce the velocity ratio produced by such drives.

Gears

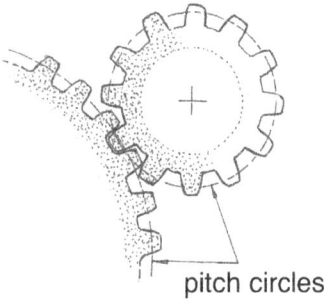

pitch circles

When two gear wheels mesh, they provide a reliable and constant velocity ratio between their respective shafts. If the teeth are worn, there might be a short backlash when the drive gear changes direction, but there can be no change to the velocity ratio between the two as a result of wear. This type of drive is called a positive drive, because there is no slip.

The pitch circle of a gearwheel is the effective circle of contact with the pitch circle of its meshing partner.

There is another kind of drive, called a friction drive, in which one rubber-rimmed toothless wheel drives another. For the purposes of the present analysis, gears may be considered to be similar to such drives. However, friction drives are inclined to slip when transmitting torque, so there is a limit to the comparison between them and gears. Friction drives are used mainly in light duty applications, where smoothness of operation and low noise levels are essential.

Assume the pitch circles of two meshing gears are equivalent to the circumferences of two friction wheels in contact, revolving without slip.

Let the angular velocity of the smaller wheel, with radius r_A, be ω_A. We need to express the angular velocity of the larger wheel in terms of this velocity.

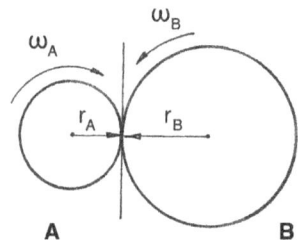

Consider a point on the rim of wheel A, at the point of contact with wheel B. This point has linear velocity $v_A = r_A \omega_A$.

Since the meshing wheels do not slip, the contact point on the rim of wheel B must have exactly the same 'linear' velocity as the contact point on the rim of wheel A.

58

It may help to imagine a piece of paper being drawn through the point of contact, and passing without slipping between the two meshing friction wheels. This 'linear' velocity would be the velocity of that piece of paper.

Now, $v_A = r_A \omega_A$, and $v_B = r_B \omega_B$ So, if $v_A = v_B$, then $r_A \omega_A$ must equal $r_B \omega_B$

If $r_A \omega_A = r_B \omega_B$ it follows that $\omega_B / \omega_A = r_A / r_B$

Therefore, the angular velocities of the two wheels are in inverse proportion to their radii. And, consequently, also in inverse proportion to their diameters, ϕ.

Example

A gear with pitch circle diameter 100 mm drives another meshing gear with pitch circle diameter 240 mm. If the smaller gear rotates at 400 r/min, what is the angular velocity of the larger gear?

$\omega_{large} / \omega_{small} = \phi_{small} / \phi_{large}$ $\therefore \omega_{large} = (100/240)(400) = 166.7$ r/min

Pitch circle ratios related to gear tooth ratios

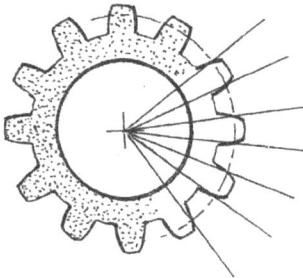

The pitch circles are the effective circles of contact for two meshing gears. If the gears mesh, the thickness of each tooth along the pitch circle must be the same on both wheels. Also, there must be an integral number of gear teeth on each wheel.

Since the pitch circles consist of whole numbers of these identical matching teeth, the number of teeth on two meshing gearwheels is in proportion to their pitch circle circumferences.

Hence, the numbers of teeth of two meshing gears are in direct proportion to the diameters (and consequently also to the radii) of their pitch circles. If the number of teeth is designated as N, then:

$N_A / N_B = \phi_A / \phi_B = r_A / r_B$

This fact allows us to use the ratios of numbers of teeth, where we know these values, instead of having to use the ratios of pitch circle diameters, which may sometimes not be known.

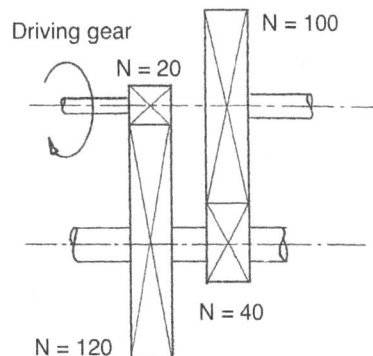

Example

A gear train consists of a driving gear with 20 teeth, which meshes with the larger gear of a compound gear (namely, two gears fixed to the same shaft) of 120 teeth. The smaller gear of this compound set has 40 teeth, and meshes in turn with the driven gear that has 100 teeth. If the driving gear rotates at 1500 r/min, determine the angular velocity of the driven shaft.

The angular velocity of the compound gear will be in inverse proportion to the numbers of teeth on the first two meshing gearwheels, namely the driving gear and the large gear on the compound gearwheel.

Hence, if the driving gear rotates at 1500 r/min, the compound gear will rotate at

1500 × (20 ÷ 120) = 250 r/min

Since the two gearwheels of the compound gear are fixed to the same shaft, they both have the same angular velocity. Therefore the smaller wheel of the compound gear is also turning at 250 r/min.

Since it meshes with the driven gear, the angular velocity of the driven gear is in inverse proportion to the numbers of teeth on these two meshing gearwheels. The angular velocity of the driven gear is thus 250 r/min × (40 ÷ 100) = 100 r/min.

The overall ratio of this gear train is 15, since it reduces an input angular velocity of 1500 units to an output angular velocity of 100 units.

Drive systems consisting of chains and sprockets, or pulleys and belts

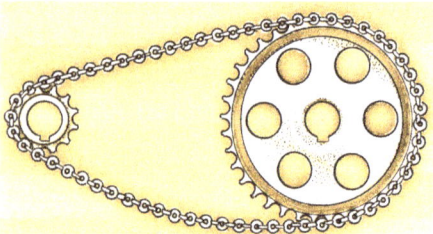

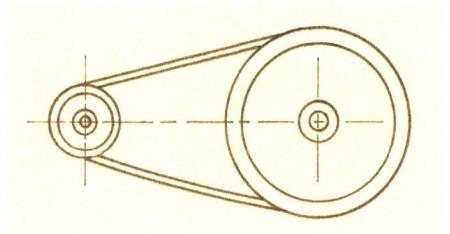

The reasoning by which we deduce a velocity ratio is the same for both these types of drive. Clearly, a drive chain does not slip relative to the sprocket that engages it. For the present analysis, we have to assume that no slip occurs on belt-driven pulleys, either. In practice, on *all* belt drives except those using toothed belts, some slip is inevitable. However, for a first approximation of the velocity ratio, the reasoning given below is valid.

Consider two pulleys connected by a rubber belt. If the driving pulley (effective radius r_A) rotates with angular velocity ω_A, we need to determine the angular velocity of the driven pulley, B.

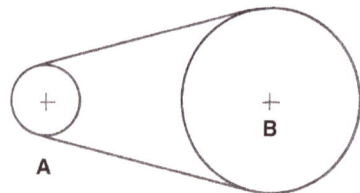

If the belt does not slip on the pulley, then the linear velocity of a point on the belt is the same as the linear velocity of the rim of the pulley, namely $v_A = r_A \omega_A$.

However, this point on the belt does not slip relative to the surface of the driven pulley, either. So, its velocity is the same, by the same reasoning, as that of a point on the rim of pulley B.

Hence, $r_A\omega_A = r_B\omega_B$, which is the same result that we had for two friction wheels in contact. So the velocity ratio between two pulleys in a belt drive is also in inverse proportion to their radii, and consequently, also to their diameters.

Even when a tensioner pulley, C, (sometimes called an idler pulley) is used, the velocity ratio between the driver and the driven pulleys remains unchanged. The introduction of an idler pulley does not affect that ratio.

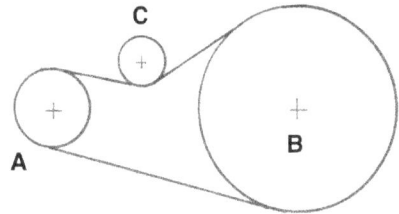

Example

A driving pulley of effective diameter 160 mm, rotating at 240 r/min, is connected by a driving belt to a driven pulley of effective diameter 560 mm. What will be the angular velocity of the driven pulley, if no slip occurs?

The velocity ratio is in inverse proportion to the diameters, so the driven pulley will have an angular velocity of (160 ÷ 560) × 240, namely 68.57 r/min.

Angular acceleration ratios in drive systems

Suppose a driving gear, A, of radius r_A accelerates uniformly from rest to an angular velocity of ω_A in t_1 seconds. Its angular acceleration will be α_A, where , applying the equation: $\omega_2 = \omega_1 + \alpha t$ we have $\omega_A = 0 + \alpha_A t_1$ so that $\alpha_A = \omega_A/t_1$.

The driven gear, B, will reach angular velocity ω_B in the same time, having accelerated at a rate of α_B, where, by the same reasoning, $\alpha_B = \omega_B/t_1$.

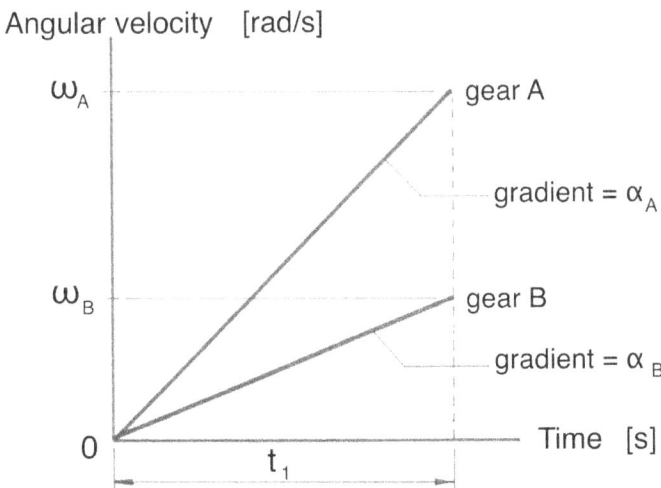

The motion of the two meshing gearwheels during this acceleration period can be illustrated on an angular velocity-time graph:

The ratio of the two accelerations is α_B/α_A

$= (\omega_B/t_1) \div (\omega_A/t_1)$

$= \omega_B/\omega_A = r_A/r_B$.

Hence the acceleration ratio of a gear train has *exactly the same value* as its velocity ratio.

Example

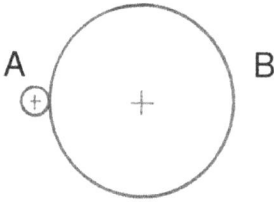

Driving gear A, with 40 teeth, engages with driven gear B, that has 280 teeth. If gear B needs to experience an angular acceleration of 2 rad/s², what should be the angular acceleration of gear A?

The acceleration of gear A must be (280 ÷ 40) times the acceleration of gear B, namely 14 rad/s².

Example

A motor car is placed on a stand so that its driving wheels are free to turn in the air. The engine accelerates uniformly from rest to 600 r/min in 4 seconds, with the wheels engaged in the drive train. The car is in first gear, where the overall gear ratio is 6.8:1. The diameter of the wheels is 700 mm. Determine:

The angular acceleration of the engine,

The angular acceleration of the wheels, and

The 'linear' velocity of a point on the rim of the wheels at the end of the four seconds.

600 r/min = 600 (2π/60) rad/s = 62.83 rad/s

$\omega_2 = \omega_1 + \alpha t$ ∴ 62.83 = 0 + α(4) ∴ α = 62.83/4 = 15.71 rad/s²

The engine will rotate 6.8 times as fast as the wheels in this gear. Since the acceleration ratio is the same as the velocity ratio, the wheels will accelerate 6.8 times as slowly as the engine. So the angular acceleration of the wheels is 15.71/6.8 = 2.310 rad/s².

The angular velocity of the wheels will be 62.83 ÷ 6.8 = 9.240 rad/s

For a point on the rim of the wheel, v = rω = 0.35(9.240) = 3.234 m/s

Exercise

A compound gear train has a velocity ratio of 16, which means the input gear turns 16 times for each turn of the output gear. If the motor turning the input gear accelerates from rest to its operating speed of 1500 r/min in 2.5 seconds, determine:

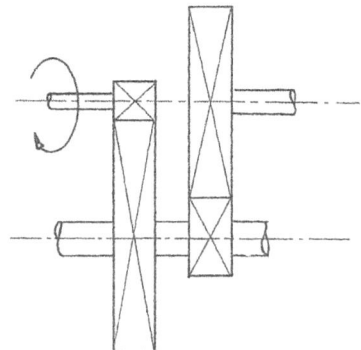

- the angular acceleration of the output gear. [3.927 rad/s²]

- the number of revolutions the output gear makes in the first 2 seconds of motion. [1.250]

- how long it will take from start-up to the moment when the output gear has rotated 10 complete revolutions, if the system continues at constant velocity after reaching operating speed. [7.650 s]

Additional exercises on all aspects of angular measurement and angular motion

Question 1

If the Earth's average radius is 6371 km, determine the shortest distance, measured along the surface, from the Equator to a point that has latitude 24°N.   [2669 km]

Question 2

Planet X orbits its sun once in 453 Earth days. Assume a circular orbit, with a radius of 217.4 million km. With what speed is this planet moving in the path of its orbit?

[34.90 km/s]

Question 3

The archway of a stone building occupies one quarter of a circle, of inner radius 4 m. If the space needed for the underside of each stone in that arch is 120 mm, how many stones are needed to form the arch? [52]

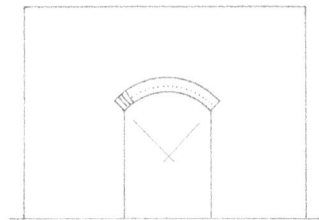

Question 4

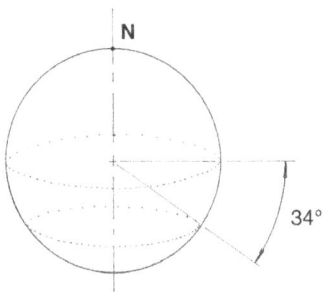

What is the linear speed of a point on the surface of the Earth, situated at latitude 34° S?   [1389 km/h]

34°

Question 5

Draw a graph of angular velocity, ω, vs. time, t, for the following motion of a drum which rotates on its axis:

a. Starting from rest, it accelerates at 4 rad/s² and keeps this up for 12 seconds,

b. It continues at constant angular velocity for a further 8 seconds,

c. It slows down uniformly to come to rest in a further 24 seconds.

From the geometry of the graph, deduce the number of revolutions the drum performed in that time.   [198.6 revs]

Question 6

A circular skating rink has diameter 60 m. If you skate along the perimeter, keeping 3 m inside the barrier fence for safety, and you take 12 seconds to complete one circuit, what is your skating speed?   [14.14 m/s]

Question 7

A bicycle rear wheel has an effective diameter of 660 mm.

*   If you rode this bicycle a distance of 2 km, how many revolutions would the wheel have made?   [482.3]
*   Through how many radians would it turn?   [3030]
*   If it takes you a distance of 60 m to accelerate uniformly up to cruising speed, and another 40 m to slow down uniformly to rest at the end of the ride, and the whole ride takes 8 minutes and 42 seconds, what is your cruising speed, in km/h?   [14.48 km/h]

Question 8

A wheel of diameter 500 mm is rotating at 34 rad/s. At time t = 0, a brake is applied, causing the wheel to slow down uniformly to 10 rad/s by the time t = 12 seconds. The wheel is then speeded up with uniform acceleration to 40 rad/s over a further 5 seconds, after which it continues at constant angular velocity, until t = 32 seconds.

a. Draw (to scale) a graph of angular velocity vs. time for this pattern of motion.
b. From the dimensions of the graph, determine the number of revolutions through which it turned between t = 0 and t = 22 seconds.   [93.74 revs]
c. Determine the 'linear' speed of a point on the rim of the wheel at t = 15 seconds. [7.00 m/s]
d. What is the value of the angular acceleration between t = 12 and t = 17 seconds? [6.00 rad/s^2]

Question 9

A rubber drive belt transmits power from a driving pulley, A (ϕ160 mm) to a driven pulley, C (ϕ580 mm). The driving pulley runs at 1200 r/min. The belt passes over an idler pulley, B, of diameter 100 mm.

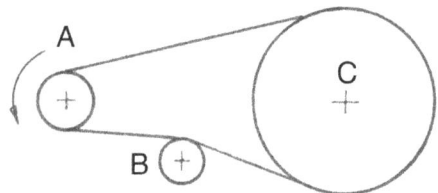

The arrangement operates like this for 2000 hours. During this time:

a. How many times has a given point on the belt passed the idler pulley? [36.2 million times]
b. Through how many revolutions have the two active pulleys turned? [144 million and 39.72 million revolutions, respectively]

Question 10

A roundabout in a children's park has radius 4 m. A child accelerates the roundabout, from rest, by running while holding an upright bar on the roundabout, and jumping onto it when she reaches her maximum speed of 5 m/s. It takes her 10 seconds from starting, to reaching her maximum speed. Once she is on the roundabout, it continues rotating while slowing down gradually over the next 60 seconds, until a point on its perimeter reaches a speed of 0.5 m/s, at which moment the child jumps off. Determine:

a. The number of radians turned by the roundabout while being accelerated [6.25 rad], and while the child was on it [41.25 rad],

b. The value of the angular deceleration undergone by the roundabout while slowing down. [0.01875 rad/s^2] and

c. The number of revolutions turned by the roundabout altogether, between starting to revolve and coming to rest. Assume that the change in the value of the deceleration of the roundabout, caused by the child jumping off, is negligible. [7.626 revs]

Question 11

Three parallel shafts are mounted on bearings (not shown) and are free to rotate with negligible friction.

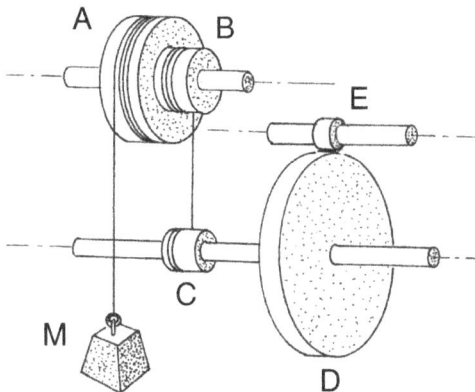

Mass-piece M is attached to a cord that may be regarded as inextensible. This cord is wound around the larger drum, A, of compound drum AB. A similar cord is wound around drum B, in the opposite direction.

The other end of the second cord is wound around drum C, which is fixed to its shaft.

Also fixed to this shaft is wheel D, which is a rubber-rimmed friction wheel that engages with friction wheel E. Assume no slip between D and E.

The mass-piece is allowed to descend from rest, and is found to descend a distance of 2 metres in 8 seconds.

The diameters are: A: 720 mm; B: 300 mm; C: 100 mm; D: 900 mm; E: 60 mm

Determine:

a. The angular acceleration of drum A. [0.1736 rad/s^2]

b. The angular acceleration of drum C. [0.5208 rad/s^2]

c. The maximum linear speed reached by a point on the rim of wheel D. [1.875 m/s]

d. The maximum angular velocity reached by wheel E. [596.8 r/min]

True/False Questions on Rotational Motion

1	A radian is a naturally-occurring measure of angle.	T	F
2	If the radius of the Earth is 6371 km, then two points on the equator that are 6371 km apart on the surface will subtend an angle of one radian at the Earth's axis.	T	F
3	The amount of angle moved through by a rotating object in a given period of time is always identical with the angular displacement of that object from the start to the finish of that period.	T	F
4	To convert rad/s to rev/min, multiply by $2\pi/60$.	T	F
5	The tangential acceleration of a point on the rim of a wheel is directly proportional to the angular acceleration of the wheel.	T	F
6	The equations of motion that relate the rotational displacement, velocity and acceleration of a rotating object have the same form as the equations which relate their linear counterparts.	T	F
7	If an angular motion is plotted on a graph of angular velocity vs. time, the area under the graph represents angular acceleration.	T	F
8	On a graph of angular velocity vs. time, it is impossible for any part of the graph to lie below the time axis.	T	F
9	If you can run at 6 m/s while holding onto the periphery of a children's roundabout of diameter 4m, you can impart an angular velocity of 3 rad/s to the roundabout.	T	F
10	For two gears A and B that mesh, $\omega_A/\omega_B = r_A/r_B$	T	F
11	The velocity ratios of gears, belt drives and chain drives are all determined by the same relation between the dimensions of the driving and driven wheels.	T	F
12	If a gearbox has a velocity ratio of 8, it follows that the output shaft will accelerate at 1/8 of the value of the acceleration of the input shaft.	T	F

Chapter 13

Work, energy, and power

Mechanical work done by a force
Work done against a force
Mechanical forms of energy
The law of conservation of energy
Power transmitted by a force
Work done by a torque
Power transmitted by a torque
Energy in gradual, sudden and shock loading

Mechanical work done by a force

'Mechanical work' must be distinguished from the everyday term 'work', which can include such diverse occupations as: entering data on a computer, organising sales by telephone, or deciphering an ancient language. None of these activities involve much mechanical work. Very specifically:

Mechanical work is the amount of energy expended by a force when that force moves its point of application through a specified distance in its own direction.

When mechanical work is performed, one or more of the following outcomes will be evident:

- A load has been moved, raised or accelerated.
- An object has been made to rotate, overcoming resistances to its rotation.
- An object has been deformed, whether temporarily or permanently.

A formal definition of mechanical work: If a continuous and non-varying force '**F**' displaces an object by a distance '**s**' along a flat horizontal surface:

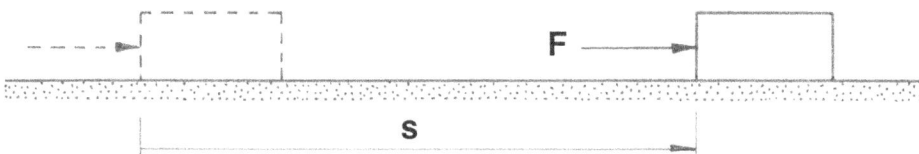

this force is said to have accomplished an amount of work done, defined by the equation:

W.D. = Fs where **F** is the force and **s** is the displacement in the direction of the force.

The units of mechanical work are equivalent to [the units of force] multiplied by [the units of distance], namely n*ewton-metres* [N.m] also known as *joules* [J].

On a graph of force vs. displacement for a nonvarying force, the area under the graph represents the work done, in joules.

(We will shortly deal with work done by forces whose magnitudes *do* vary).

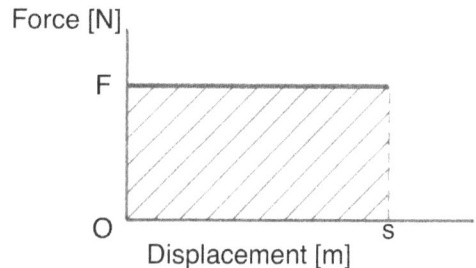

Sometimes a force *acts*, but does *not* perform work

If, due to resistances experienced, a force is not succeeding at displacing its point of application, *no* mechanical work is being done.

For instance, you could hold a 50 kg load steadily above your head for as long as you were capable of doing so, but while the load is in that position, you are not performing mechanical work on it. When you raised the load to that position in the first place, you exerted a force through a distance, and therefore did work on the load. It will feel as if you are expending a lot of energy in holding the load above your head, but you are not doing any further mechanical work *on the load.*

Similarly, if two tug-of-war teams pull their hardest, but neither team can budge the other, no mechanical work is being done.

Typical situations in which mechanical work is performed

1. A hand-saw is driven through wood,
2. Water is pumped uphill to a storage dam.
3. A load is trucked up a mountain road.
4. The ball of a demolition crane smashes into a wall, crushing part of the wall.
5. Expanding gases push a piston down a cylinder.
6. An archer draws back a bow-string to shoot an arrow.

In many of these instances, the value of the force that is doing the work does not remain constant over time. In reality, a situation seldom occurs in which a force maintains a constant value while performing work.

Work performed by a force of varying value

One example of a force which varies while it is performing work, is the force required to compress or extend a coil spring. The further you compress a spring, the greater the applied force needs to be. This relation is linear, in the elastic range of a spring, which is that range of extension for which the spring returns to its original shape once the force is removed.

The linear relation between force and extension in a spring, known as Hooke's Law, was first described by Robert Hooke (1635 - 1703) who was at one time Curator of Experiments at The Royal Society, and a scientific investigator of note, with many accomplishments to his name, in a variety of fields of research.

The following graph shows force F vs. compression x, for a coil spring behaving elastically.

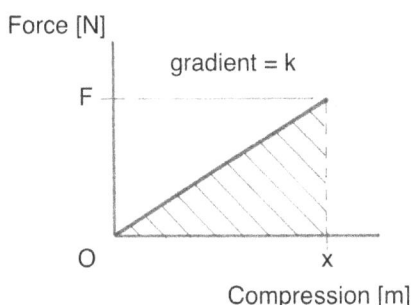

As in the case of a non-varying force, the amount of work done is also represented by the area under the graph.

If the gradient of the graph is k, then $F = kx$.

Thus the area under the graph up to a position 'x' on the horizontal axis is a triangle, with area given by ½(base × height), namely ½ $(x)(kx) = $ ½ kx^2.

This represents the amount of work done to compress the spring from a relaxed position (where $x = 0$), up to the point where the compression is x. That amount of work is now stored as energy in the compressed spring.

The factor k is known as the 'spring constant' or the 'stiffness coefficient' of that particular spring. The stiffer the spring, the higher the value of k. The units of k are [N/m]. For instance: if a spring has a k-value of 2400 N/m then it requires 2400 N to compress the spring by a distance of 1 m (provided the spring is long enough to be compressed by a whole metre) or 2.4 N to compress it by 1 mm.

If a force varies non-linearly while it is performing work, the amount of work done is still represented by the total area under the graph of force vs. displacement.

An example of such a varying force would be the force transmitted to an arrow by a bowstring, as represented here:

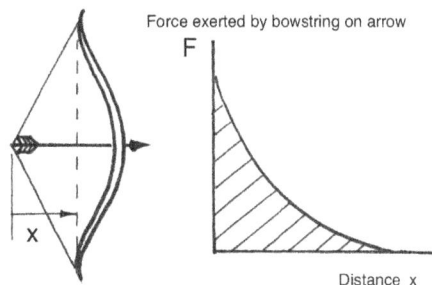

Example

How much energy is stored in a coil spring with stiffness k = 4500 N/m when the spring is compressed by 20 mm from a relaxed position?

Energy stored in the spring = work done on the spring

= ½ kx^2 = ½ $(4500)(0.02)^2 = 0.9$ J

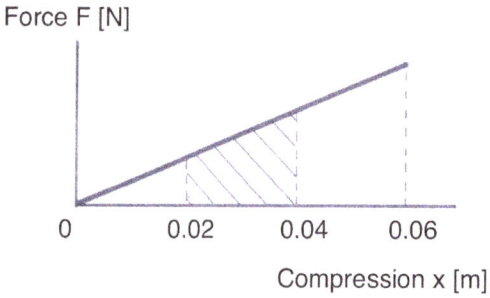

Force F [N]

0   0.02   0.04   0.06

Compression x [m]

How much work must be done to compress this same spring from an initial compression of 20 mm to one of 40 mm?

The work done is represented by the area under the graph between

x = 0.02 m and x = 0.04 m, namely:

½(4500)(0.04² − 0.02²) = 2.7 J

Similarly, the work done to compress this same spring from x = 40 mm to x = 60 mm would be 4.5 J.

Work done by an oblique force

If the movement caused by a force takes place in some line of action at an angle to the line of action of the force, then, in determining the work done, *only* the distance moved parallel to the line of action of the force is counted.

Example

A horizontal force of 436 N is needed to push a trolley at slow constant velocity up a ramp making an angle of 25° with the horizontal, for a distance of 16 m. How much work is done by this force?

The distance moved parallel to the line of action of the force is 16 cos 25° = 14.50 m.

Work done by this force is
436 N × 14.50 m = 6322 J

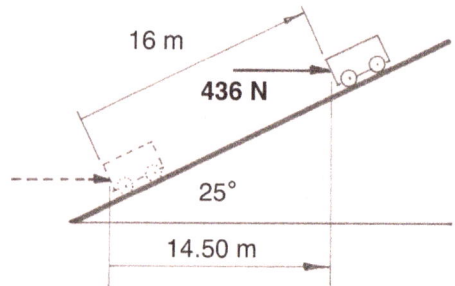

16 m

436 N

25°

14.50 m

Example

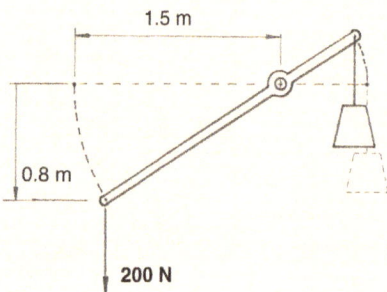

1.5 m

0.8 m

200 N

In order to raise this particular load, one has to pull vertically downwards on a rope attached to a lever, with a force of 200 N.

How much work would you do in moving the point of application of the rope downwards by 0.8 m?

The point of application of the rope moves in an arc, whose length is clearly greater than 0.8 m.

However, the force has moved only 0.8 m in a direction parallel to its line of action, and it is therefore *this* displacement which is used to determine the amount of work done, which is 200 N × 0.8 m = 160 J.

Work done 'against' a force

Recall the zeroth law of mechanics: a force cannot even exist unless it is opposed.

Imagine an astronaut floating in space, unattached to any other object. No matter how strong he is, there is no possibility that he can perform any mechanical work on his surroundings, because while he is floating freely like that, there is nothing around him to provide a resistance to his effort.

However, if he were to push against the side of a space station, he could exert a force through a distance, and thereby perform mechanical work. The effect of his pushing against the space station would be to accelerate both him and the space station away from each other.

He could also exert a force against a resistance if he were holding, for example, a flexible rod in his hands, and could bend the rod. In this case he would be exerting a force against the resistance that the material offers to bending. By bending the rod, he would be performing mechanical work.

We exert a force in a mechanical situation to overcome other forces in order to achieve an effect we want. For instance, we may need to overcome forces of:

- Gravity, as when raising an object
- Friction, as when dragging an object
- Structural or material strength, as when bending a bar
- Inertia, as when accelerating an object

By pushing back a force F a distance 's' along its own line of action, we do work W.D. = Fs *against* the force.

- When a force raises an object in a gravitational field, it performs work *against* gravity.
- When a force drags an object while opposed by friction, it does work *against* friction.
- When a force succeeds in deforming material, it does work *against* the resistance of the material.
- When a force succeeds in accelerating an object, work is done *against* inertia.

Example

How much energy is lost to friction when a 20 kg stone block is dragged a distance of 5 m along a horizontal concrete floor, given that $\mu = 0.7$ between these materials, and that the kinetic coefficient of friction is 75% of the static one?

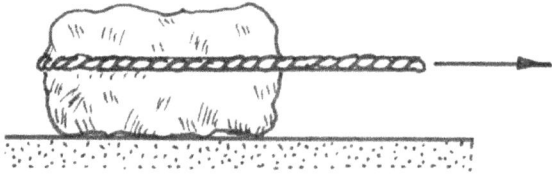

While the block is stationary, you have to exert sufficient force to overcome the limiting friction force, F_{max}. However, while there is no movement, no work is being done. Work is done only when the force exerted by the rope has moved its point of application. Once sliding commences, the applied force required to keep the block sliding must be large enough to overcome kinetic friction.

The magnitude of this force is $F_k = \mu_k N$ where N is the normal reaction between the two sliding surfaces, which in this case is equal to the weight of the block.

$\therefore F_k = (0.75)(0.7)\,(20 \times 9.81) = 103.0$ N

$\therefore$ the work done against the friction force is equal to (the force × the distance by which this force has been pushed back in its own line of action) namely = 103.0 × 5 = 515.0 J, so this amount of energy is the friction loss.

Performing *more* work than that required to overcome the resistances

When a force P overcomes a force F which is resisting it, we say P does work *against* F.

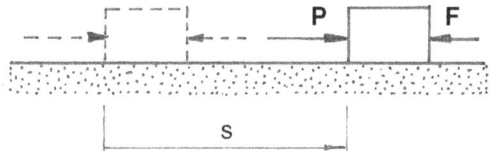

In order to overcome force F, force P has to be at least marginally greater than F.

If P is substantially greater than F, then the extra work done, namely (P − F)s , results in an additional outcome such as acceleration or deformation of material.

For example, suppose it requires a force of 100 N to push a stone block along a horizontal surface against the friction force, with slow constant velocity. If we exert 110 N instead of 100 N, the block will not only slide, but will also accelerate.

The force available to produce acceleration is 110 − 100 = 10 N. The magnitude of the resulting acceleration will be given by a = F/m = (10 N)/(20 kg) = 0.5 m/s².

Exercises on work done by a non-varying force

In questions 1 and 2 that follow: determine the work done by the effort force.

1. A motor raises a passenger lift (elevator) that weighs 12 kN , through a height of 75 m, at constant velocity. There is friction in the guides, assumed constant at 400 N. [930 kJ]

2. A plough horse drags a plough to make a furrow that is 50 m long. The force needed to cut the furrow is 600 N. [30 kJ]

3. A 50 kg wooden crate is dragged a distance of 8 m up a wooden ramp that makes an angle of 15° with the horizontal. If the coefficient of static friction is 0.6 and the coefficient of kinetic friction is 80% of that:

- How much force applied parallel to the ramp is needed to make the crate move?   [354.4 N]
- How much energy is expended by the applied force,   [2835 J] and
- How much energy is lost to friction?   [1819 J]

Work done when the force performing work varies during the motion

Example 1

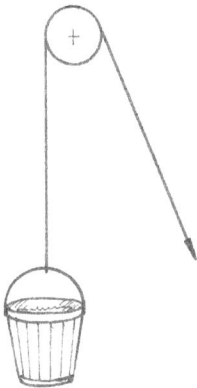

Water is raised from a well by a wooden bucket and chain, passing over a sheave. The mass of the bucket is 8 kg and that of the water it contains is 16 kg. The mass of the chain is 0.5 kg/m. Friction in the sheave adds an effective 2 kg to the load. Assume that the length of chain between the sheave and the operator remains constant.

Determine the mechanical work done by a person to raise this bucket a height of 6 m.

The variation in force from the start to the finish of the motion is caused by the diminishing length of chain that has to be raised.

When the bucket is at its lowest point, the mass of chain to be raised is 3.0 kg, and when at its highest point, zero.

So, the maximum force needed is $(8 + 16 + 3 + 2)\,9.81 = 284.49$ N

And the minimum force needed is
$(8 + 16 + 0 + 2)\,9.81 = 255.06$ N

The work done is represented by the area under the graph of force vs. displacement.

Since this area is a trapezium, its value is equal to (base) × (average height), namely:

$6 \times (½\,(284.49 + 255.06)) = 1619$ J.

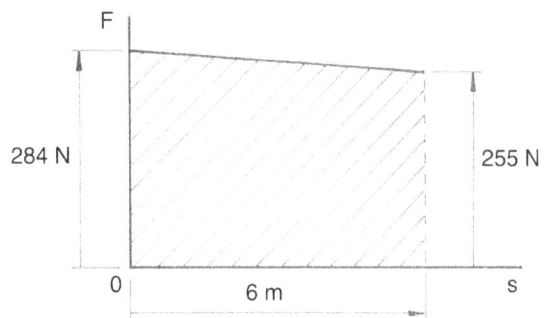

Example 2

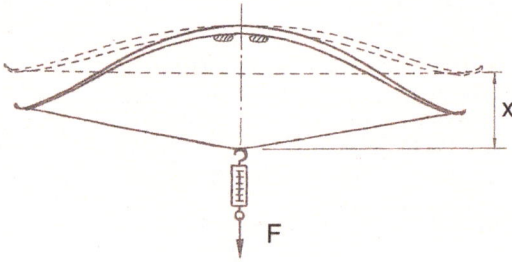

An archery bow is suspended over two fixed supports. The bow-string is extended gradually, by a downward force F that can be varied and measured.

The values of force F and extension x are given in the table below:

Extension, x [mm]	0	200	400	600	800
Force F, [N]	0	20	46	90	160

Determine the approximate amount of energy stored in this bow at full draw, namely when x = 800 mm.

Since the amount of work done on the bow-string becomes energy stored in the bow, we need to determine the area below the graph of F vs. x.

Draw a graph showing force vs. extension, to scale. Although the graph is a curve, it can be approximated to consist of four vertical bars in the shape of trapeziums. The area of each one can be determined as (base × average height).

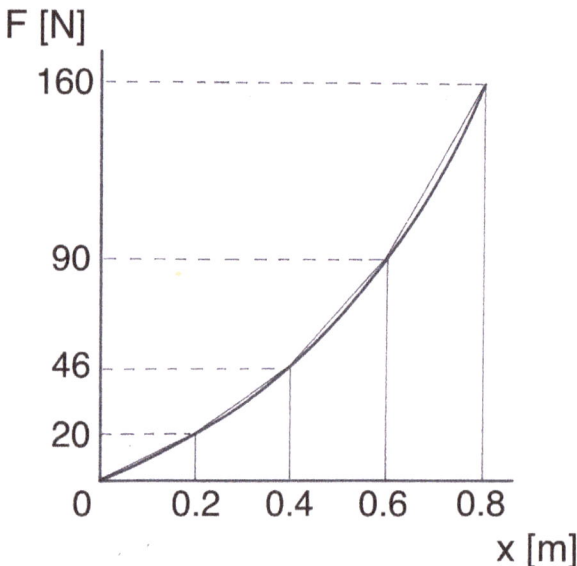

Areas of the respective bars:

1st bar: 0.2(0 + 20)/2 = 2.0 J

2nd bar: 0.2(20 + 46)/2 = 6.6 J

3rd bar: 0.2(46 + 90)/2 = 13.6 J

4th bar: 0.2(90 + 160)/2 = 25.0 J

Total area = 47.2 J.

Final estimate is approximately 46 to 47 J, since each trapezium contains slightly more area than the area under the plotted curve between its extremities.

Exercises on work done by a varying force

Question 1

When a mass-piece weighing 100 N is suspended from a coil spring, it extends the spring by 8 mm. Determine:

The stiffness coefficient of the spring. [12500 N/m]

The mass of a block that would extend the spring by 25 mm when suspended from it. [31.86 kg]

The work done on the spring to stretch it by 25 mm from its un-stretched state. [3.906 J]

The additional work done on the spring in extending it from an extension of 25 mm to one of 50 mm. [11.72 J]

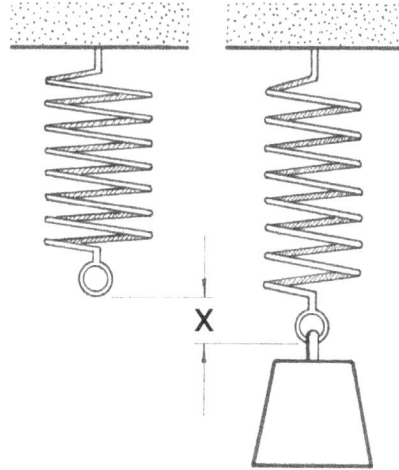

Question 2

A lift (elevator) of mass 804 kg is raised a height of 42 m at slow constant velocity by a winch driven by an electric motor. The friction in the lift guides is 412 N. The wire rope raising the lift has mass 2.8 kg/m. The combined mass of the people in the lift is 298 kg.

Determine the total amount of work done by the motor. [495.6 kJ]

Question 3

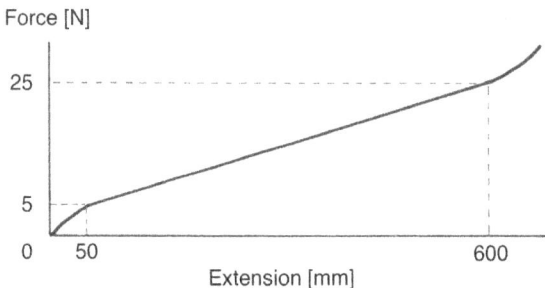

A rubber band produces a force vs. extension graph approximately similar to that of a coil spring, in the mid-range, where it behaves elastically.

For the rubber band whose characteristics are described on the accompanying graph:

How much energy is stored in the rubber band by stretching it from $x = 50$ mm to $x = 600$ mm? [8.25 J]

How much energy could be stored in 12 such rubber bands, acting in parallel, if they were stretched from $x = 50$ mm to $x = 400$ mm? [50.4 J]

Question 4

A digging bucket weighing 40 N is pulled along a horizontal stretch of beach sand at slow constant velocity. The coefficient of friction between the bucket and the beach is 0.6.

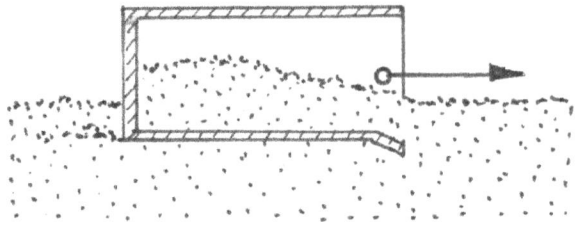

After being pulled a distance of 5 m, the bucket, with the sand accumulated in it, now weighs a total of 180 N. How much work was done to fill the bucket? Assume the mass of the sand in the bucket increased linearly over the five metres.   [370 J]

Forms of Mechanical Energy

Energy is the stored-up ability to perform mechanical work.

Sources of energy for mechanical applications include: direct heat from combustion, electrical charge, chemical energy, nuclear energy, solar energy, and energy from the motion of wind, waves and tides. In order to make efficient use of these sources, it is convenient to store the energy for use when required.

Appreciable amounts of energy can be stored in an electric grid, or in the form of chemical energy, such as in batteries, or by forming compounds that can release energy in a reaction (like fuel or explosives). However, energy can be stored *mechanically* in only three forms, and the amount of energy that can be stored in a mechanical form is generally much smaller than that which can be stored in the other forms mentioned above.

The three mechanical forms of energy are:

- Gravitational potential energy;
- Kinetic energy; and
- Energy stored in a stretched, flexed or compressed elastic material, such as a spring.

When analysing mechanical interactions, we make use of the fact that mechanical energy can change from one form to another in certain circumstances, allowing us to determine values of interest to a given problem, such as velocities attained, heights reached, and forces exerted.

Before looking at the way mechanical energy changes from one form to another, we will define the three forms of mechanical energy and show how the equations for their magnitudes are derived.

Gravitational Potential Energy (PE)

Usually known simply as 'potential energy'. This form of energy arises because a mass that descends in a gravitational field has the potential to perform mechanical work (namely, to exert a force through a distance) equivalent to the work that was done on it to raise it to the original elevation.

Examples of PE

1. Water in a storage dam can flow to a lower altitude, where it can be used to drive a turbine to generate electricity, or to drive a machine directly, through a water-wheel.

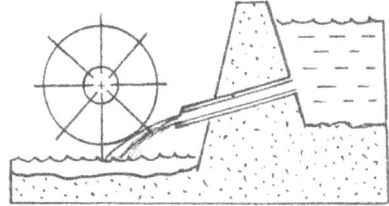

2. A raised mass-piece can be attached to a rope passing over a sheave and wound around a drum. The descent of the mass-piece can release energy to rotate the drum and thus perform work.

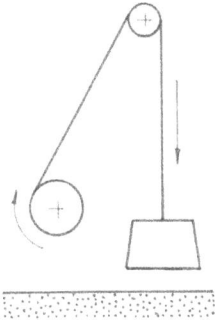

3. Any mass can be dropped, swung, rolled or allowed to slide down from a height, and will exert a force where it strikes, capable of performing work, thereby causing motion in other objects, or deformation of material.

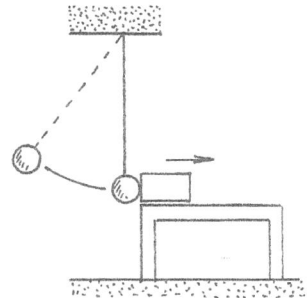

The expression for PE

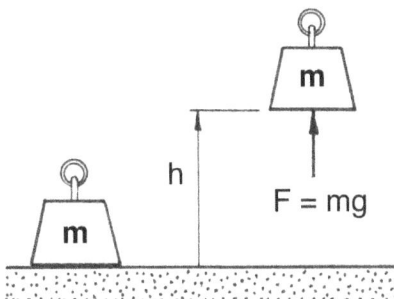

The amount of gravitational potential energy possessed by a mass in a gravitational field is defined by the expression $PE = mgh$

where m = the mass, g = the gravitational acceleration at that place, and h = the height above a reference level chosen for the situation. The work done on an object to raise it a height 'h' above the chosen reference level is given by

(force × distance) = mg × h = mgh

When dealing with gravitational potential energy, it is essential to remember that all

77

PE is measured relative to a *reference level for zero PE*, which can be chosen to suit the needs of a given problem. The chosen reference level must be specified at the start of the solution of any problem in which changes between different forms of energy will be considered. Once you specify the reference level for zero PE, this reference level should remain the same throughout your analysis.

Kinetic Energy (KE)

A mass that is moving, whether in a straight line or rotating, has the ability to perform mechanical work (exert a force through a distance) by being slowed down. The original amount of work done to accelerate the object is now available to be done on the surroundings.

Examples of kinetic energy

1. A moving vehicle collides with an oil drum in its path, causing deformation of material and transferring some movement to the drum.

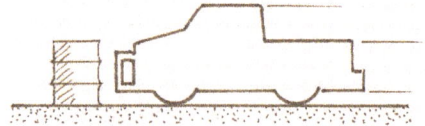

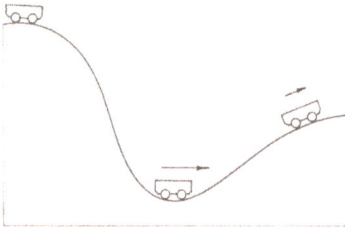

2. A roller-coaster trolley rolls down, through the bottom of a dip, and by virtue of its velocity, it still has sufficient kinetic energy to roll up the other side of the dip, thereby doing work against gravity.

3. A large grindstone wheel that is freewheeling, can still remove material from a tool tip until it has slowed down to rest.

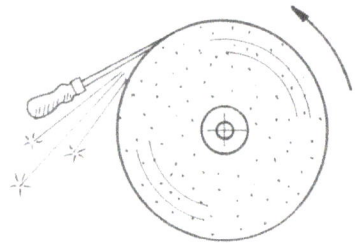

In the present chapter we will deal only with kinetic energy related to linear movement. The kinetic energy of rotating objects will be dealt with in a later chapter, once the concept of the inertia of rotating objects has been explained.

The expression for KE

The amount of kinetic energy possessed by an object of mass **m** moving with velocity **v**, is given by: $KE = \frac{1}{2}mv^2$. This expression is derived as follows:

In order to accelerate an object of mass m from rest to a velocity v, a force F must be exerted on the object. Suppose the velocity v is achieved after moving through a displacement s. The work done on the object by the accelerating force F is given by:

W.D. = Fs.

The force is related to the acceleration by F = ma, $\therefore$ a = F/m

The acceleration, a, is related to the final velocity by the equation: $v^2 = u^2 + 2as$.

In this case the initial velocity, u = 0,

$\therefore v^2 = 0^2 + 2$ (F/m)s

$\therefore$ Fs $= \frac{1}{2}$ m v^2 = work done on the object = kinetic energy possessed by the object.

Energy stored by material in elastic strain (elastic energy, or EE)

The word 'elastic' is a specific term used in mechanics, to describe a material that behaves in such a way to return to its original shape when the force causing it temporarily to change shape is removed. When a material is behaving elastically, the amount of extension or compression it experiences is directly proportional to the applied force. (Hooke's Law)

Examples of EE

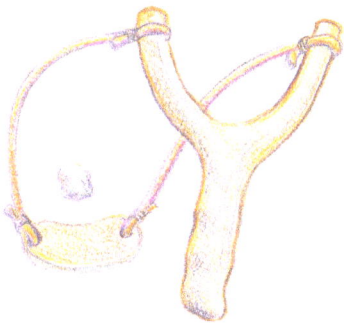

1. The work done to compress or extend a coil spring or rubber band is available to be done on its surroundings when the spring is released.

2. When an archery bow is drawn, elastic energy is stored due to the flexing of the bow. Ideally the bowstring should be inextensible, so that all the work done by the archer goes into flexing the bow, not extending the string. When the bow-string is released, most of this energy is transferred to the arrow, imparting acceleration to it.

3. A diver imparts energy to a springy diving board, in a series of jumps, until the board is springing back enough to provide additional energy for the start of the dive.

4. An air-gun works by compressing a contained volume of air. When the trigger is pulled, the pressure due to that volume of air trying to return to its original volume forces the pellet down the barrel.

5. A mechanical clock is wound up by having its mainspring rolled tightly into a compact coil. The spring tries to regain its original shape, and by unwinding provides energy to drive the mechanism.

Not all materials behave elastically. Those that do, have a limited range of movement for which they will remain elastic. If we stress a material beyond this point, it begins to deform permanently. When the stress is removed, the object may return partially to its former shape, or it may not regain its shape at all. For example, the small coil springs that are often found in ball-point pens can easily be drawn out to the point where their coils won't return to their original shape. When an applied force results in a permanent shape change, the material is said to behave *plastically*.

The expression for the amount of EE stored in a spring

We showed previously in this chapter that the work done to compress (or extend) a coil spring by a distance x is given by $\frac{1}{2} kx^2$, where k is the stiffness coefficient of that spring.

The amount of elastic energy stored in a spring is therefore: $EE = \frac{1}{2} kx^2$

The amount of elastic energy stored in other devices (such as an archery bow, air-gun, diving board, spear-gun bands and clock mainspring) will be governed by expressions that are more complicated than the above expression. We will not develop expressions here for these uniquely varying situations. However, the principle of the storage of elastic energy in these devices is the same. If we can determine how much work is done in deforming a device elastically, then that is the amount of energy is stored in the device.

Exercises on work and energy

1. A 100 kg ball is suspended from a 10 m rope. The ball is pulled to one side so that the rope makes an angle of 40° with the vertical. How much PE has the ball gained?   [2295 J]

2. A mass-piece of 125 kg, suspended from a coil spring, extends the spring by 21 mm.

   What is the value of the spring constant? [58.39 kN/m] How much mass should be added to the first mass in order to extend the spring by a total of 46 mm? [148.8 kg] How much work is done in stretching the spring by 46 mm? [61.78 J]

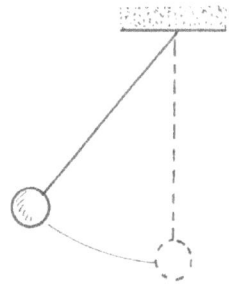

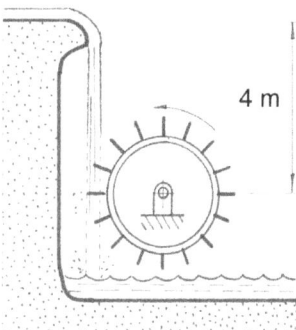

3. Water moving at a negligibly slow horizontal speed falls through a height to provide energy to turn a waterwheel. How much energy is potentially available from 1 m^3 of water falling through a height of 4 m? [39.24 kJ]

Do you think all this energy would be successfully transferred to the wheel? Why, or why not?

4. A package with mass 40 kg slides down a ramp inclined at 35° to the horizontal, for a distance of 15 m. The coefficient of kinetic friction between the package and the ramp is 0.3.
How much work does the sliding package do against friction? [1446 J]

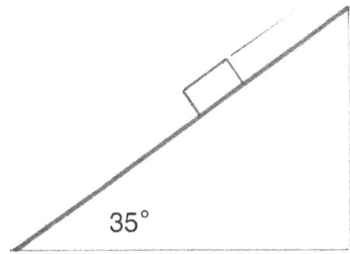

5. If you push an 800 kg car which has rolling resistance of 30 N up a slope of 2°, for a distance of 10 m, what force do you have to exert, and how much mechanical work do you perform? [303.9 N; 3039 J]

6. A 'human cannonball' of mass 70 kg is shot into the air at an angle of 45° by a spring-powered 'cannon'. The piston that his feet rest on moves through a distance of 5 m in the process of launching him. He reaches a maximum height of 12 m above the mouth of the cannon barrel. Assume he is accelerated uniformly from rest. Ignoring any energy losses due to air resistance, determine:

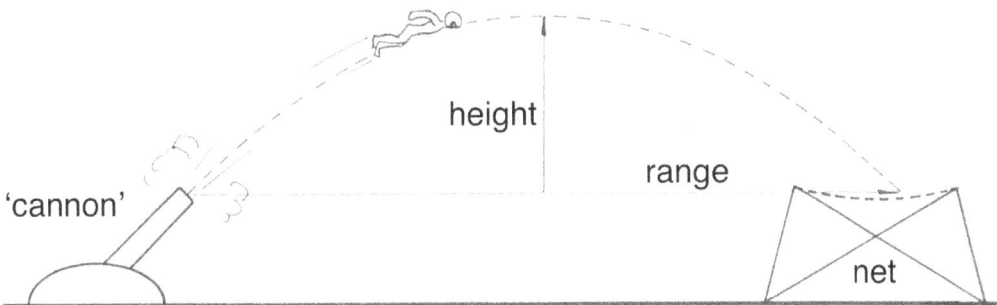

  • The vertical component of his velocity on leaving the barrel. [15.34 m/s]
  • His actual velocity on leaving the barrel.   [21.70 m/s]
  • The average force on his feet during the launch.   [3296 N]
  • The number of g's he experiences during the launch. [4.8]
  • How far he is likely to fly, if he lands on a safety net at the same height as the mouth of the barrel. [approx. 48 m]

7. The exhaust gases from a firework rocket, of mass 200g (assume this mass remains constant, even though in reality, it diminishes with the burning of the fuel) exert a propulsive force of 4 N for 5 seconds. Ignoring air resistance, if the rocket is fired vertically upwards from rest, determine:

- The upward acceleration that the rocket undergoes. [10.19 m/s^2]
- The maximum velocity reached by the rocket [50.95 m/s, approx. 183 km/h]
- The height to which the rocket rises while still under propulsion. [127.4 m]
- The work done by the exhaust gases. [509.5 J]
- The maximum height reached by the rocket. [259.7 m]

The Principle of Conservation of Energy

This principle, like all the other laws of mechanics, is based on observation. It states that the total amount of mechanical energy in a given system of bodies is *conserved*.

This means that all the energy possessed by a given set of objects, in whatever form it may be, remains *accountable*. Although energy may change from one form to another, and some might be irretrievable, due to dissipating in the form of heat, sound and deformation of material, all of it can be accounted for, because the total amount present must remain constant.

Examples describing how mechanical energy can change its form

A situation without impact:

A vehicle moving at velocity, and allowed to freewheel up a hill, will slow down and come to rest eventually. In so doing, it has gained height, and therefore potential energy. So, whereas its KE is reduced, its PE has increased.

By the Law of Conservation of Energy, the amount of PE it gains cannot exceed the amount of KE it started with.

Ideally, the amount of PE gained will be equal to the amount of KE lost. However, in practice, the amount of PE gained will be *less* than the original amount of KE, due to work having had to be done against rolling resistance and air resistance. The fact that these losses occur does not negate the principle that energy is conserved. As long as we can account for where the energy went, and how much was lost, the principle still applies, namely that the total amount is conserved.

A situation involving an impact:

A wrecking ball, used in demolishing an old building, swings from a stationary position, some height above the aiming point. It starts with an amount of PE, which, during the swing, becomes converted to KE. When it strikes the wall, its energy is consumed in work done to deform the material of the wall.

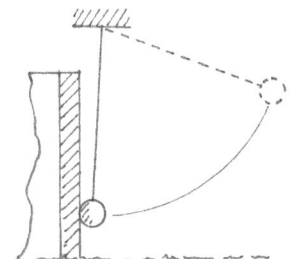

Almost all of the energy used in this way is irretrievable.

Some of it is is converted to heat and to sound. We can explain where the energy went, and we can sometimes determine how much was lost, but we can't get it back in any useful form.

Very important: In situations such as the above, in which impacts occur, the principle of conservation of energy still applies, but *cannot* be used to guide our calculations, as too great a proportion of the available energy is lost. The amount of energy lost in an impact is always significant, and in some instances can reach 100% of the original quantity of mechanical energy present.

We will see how to deal with impact situations in chapter 16 on 'linear momentum'.

When objects interact *without* impacts, it is possible to determine certain variables like velocity reached or height gained, by considering the total amount of mechanical energy in the system to be conserved.

Example:

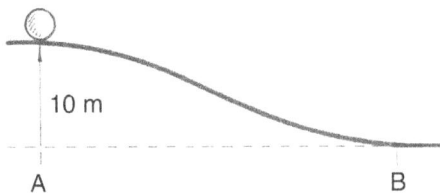

A 5 kg ball rolls from rest down a smooth continuous slope from point A through point B, a distance of 40 m down the slope. Ignoring friction, determine its velocity at point B.

Let the reference level for zero PE pass through point B.

At point A, the ball has zero KE, and its PE = mgh = $5 \times 9.81 \times 10 = 490.5$ J

Since no energy is lost or gained en route to point B, the total energy at B must equal the total energy at A. Thus the total energy of the ball at point B is also 490.5 J.

By definition, the PE of the ball, relative to the chosen reference level, is zero at point B. Therefore its total energy at point B consists entirely of kinetic energy. Suppose the ball has velocity v when at point B.

$\therefore \frac{1}{2}(5)v^2 = 490.5$ $\therefore v = 14.01$ m/s^2. This solution has disregarded any resistances that might have diminished the energy of the ball on its path from A to B. However, what happens if there are resistances, causing energy to be 'lost' along the way?

Supposing this ball experienced a rolling resistance of 1.5 N on this slope. If the total distance rolled from A to B is 40 m, then the work done against rolling resistance would be (force × distance) namely $1.5 \times 40 = 60$ J. This amount of energy is unrecoverable, hence the total mechanical energy of the ball at point B would be diminished by 60 J. Equating the initial energy and the final energy becomes:

$\frac{1}{2}(5)v^2 = 490.5 - 60$ and hence v = 13.12 m/s

In reality, any object moving in air also experiences air resistance. We will ignore air resistance in this chapter because it varies with the speed of the moving

object, and would complicate our calculations. However, it is well to note that the magnitude of air resistance is usually quite significant. The topic of air resistance is treated in more detail in the chapter on vehicle dynamics in volume 3 of this series.

Example

A 200 kg wagon is initially at rest, in position A. It rolls through the dip at B, and some way up the adjoining hill before coming to rest momentarily at point C.

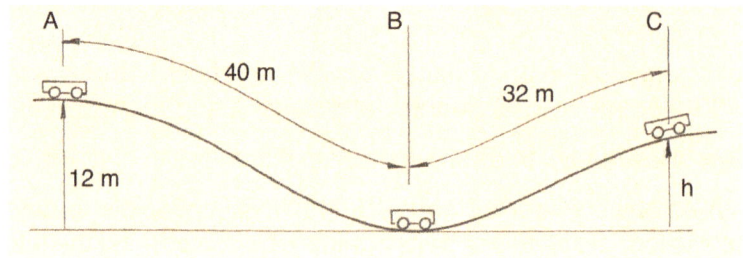

The rolling resistance of this wagon is known to be 55 N, and does not vary with speed.

Determine the velocity of the wagon at B and the height h when it reaches point C.

In this particular problem, it makes sense to choose the level for zero PE to coincide with the bottom of the dip.

In position A, all the energy possessed by the wagon is gravitational potential energy, given by: PE = mgh = 200 × 9.81 × 12 = 23544 J

This amount of energy will continue to be the total at each successive stage of the progress of the wagon, in moving from point A, through point B, to point C.

On its way to point B, some of this amount of 23544 J is 'lost' to work done against rolling resistance. The work done vs. rolling resistance while moving from A to B is (force × distance) = 55 N × 40 m = 2200 J.

The remainder of the initial energy is transformed into KE. At point B, the wagon no longer possesses any PE, because, by our definition, point B is at the level for zero PE.

The KE of the wagon at position B is 23544 - 2200 = 21344 J. To obtain the velocity:

KE = ½ mv²   ∴ 21344 = ½(200)v²   ∴ v = 14,61 m/s = approximately 52.6 km/h

The work done vs. rolling resistance between B and C is 55 N × 32 m = 1760 J

When the wagon comes to rest momentarily at point C, it has zero velocity, and thus zero KE. The KE that it possessed at B has been diminished by the work done vs. rolling resistance. The energy that remains has been transformed into PE.

The PE at position C is therefore 21344 -1760 = 19584 J. The height above the reference level reached by the wagon at this point is determined from: PE = mgh

∴ 19584 = 200 × 9.81 × h   ∴ h = 9.982 m

Energy at A	Energy at B	Energy at C
	Loss = 2200 J	Loss = 2200 J
		Loss = 1760 J
PE = 23544 J	KE = 21344 J	PE = 19584 J

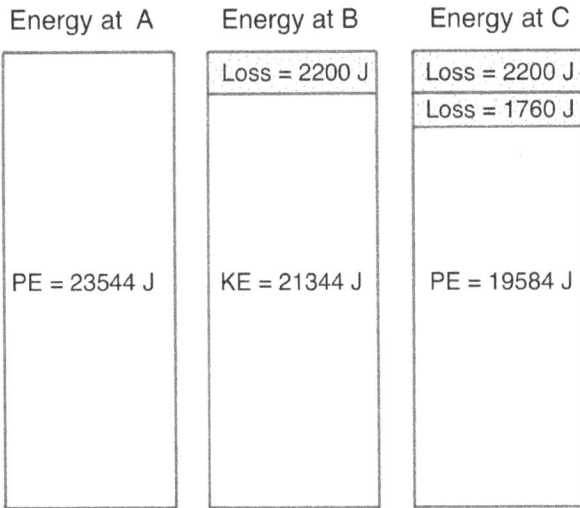

Shown here is an energy accounting diagram for the movement of this wagon from points A to C. This diagram is essentially a bar chart, drawn to scale, with each bar having the same total height, reflecting the conservation of energy from stage to stage of the movement of the wagon.

Drawing such diagrams can assist us to clarify our thinking about what is happening to the energy in a given situation.

Exercises on the conservation of energy

Question 1

A 90 kg wagon rolls down a hill, from rest, and up the side of the next hill before coming momentarily to rest again, as shown.

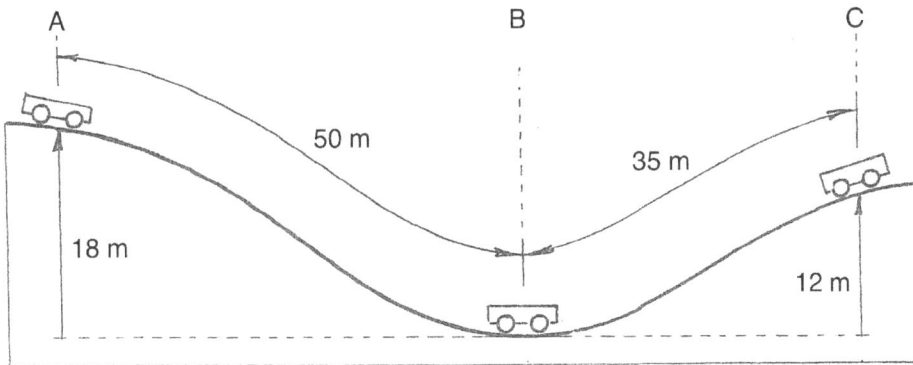

Ignoring air resistance, determine the rolling resistance of the wagon, and the velocity it reaches when at the lowest point of the dip. Illustrate the manner in which the energy changes form, from positions A to B to C, by means of an energy accounting diagram.

[62.32 N and 16.85 m/s or approximately 60.7 km/h.]

Question 2

A train of mass 400 tonnes is moving on a level track at 72 km/h, when the brakes are applied, slowing it down to 36 km/h. How much energy does the train lose? [60 MJ] What happens to this energy?

Question 3

A spring of stiffness k = 2000 N/m is contained in a vertical tube, under a bar of negligible mass that moves in a slot. The spring is relaxed when the bar is in its highest position in the slot. A ball of mass 100 grams is placed on the bar. The bar is pulled down until the spring is compressed by 20 mm, and then released, projecting the ball vertically upwards.

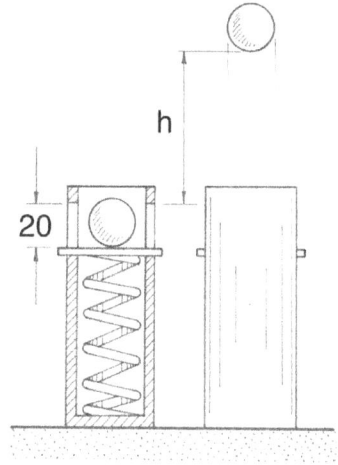

Determine:

The velocity of the ball as it leaves the bar at the top of the spring. [2.758 m/s]

The height above the projection point that the ball will rise before falling back. [0.3877 m]

Question 4

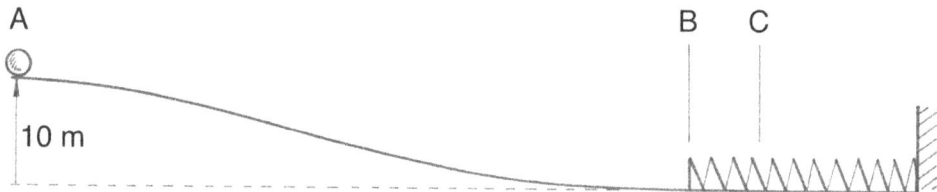

A 20 kg steel ball rolls from rest at point A, down a hard, smooth slope, which gradually becomes horizontal. The ball hits and compresses a coil spring whose extremity is originally at point B. The spring constant of this spring is 400 kN/m.

The impact compresses the spring to point C, before it rebounds. Assuming the rolling resistance of the ball is negligible, determine the velocity of the ball just before striking the spring at B. [14.01 m/s]

If the distance from B to C is 80 mm, how much energy was lost in the impact when the ball hit the spring? [682 J]

Question 5

An 8 kg steel ball rolls down a rubber-surfaced slope, from rest at point A, through a dip at point B, and up a ramp where it loses contact with the surface at point C. Point C is 0.5 m above B. The ball continues through the air and reaches the highest point of its trajectory at D before descending further.

Determine its velocity at point B, if 5% of the energy it possessed at A is lost to friction by the time it gets to B. [7.478 m/s]

If 3 % of its energy at B is lost to friction by the time it gets to C, what is its velocity at point C? [6.666 m/s]

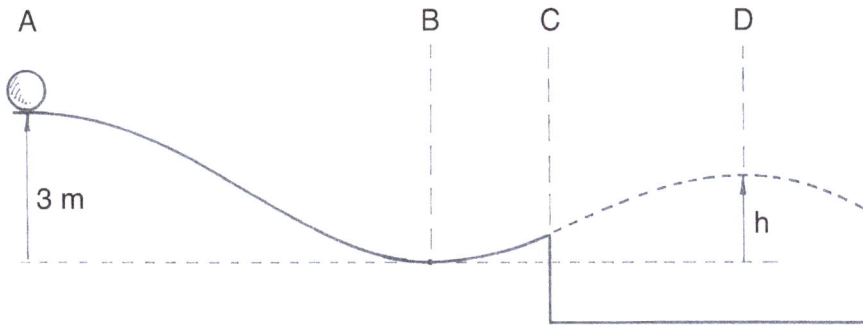

What will be the maximum height above point B, reached at point D? [0.8178 m]

Draw a set of energy-accounting diagrams to show the way the mechanical energy possessed by the ball changes its form, from positions A through to D.

Power

Power is the rate at which work is done, or energy expended.

Many of the achievements of modern engineering have been made possible by the development of machinery that can expend energy at a greater rate than humans can.

Think of digging a hole to plant a tree. You might take an hour of hard work to dig the hole using a spade, whereas a small earth-mover could remove that amount of soil in one grab.

Power = (work done) ÷ (time taken)

The units of power are:

[units of work] ÷ [units of time]
= [N.m] ÷ [s] = [J/s] = watts [W]

For example, using a block and tackle, a man might raise a weight of 10 kN through a height of 1.0 m, taking 200 seconds. The work done = Fs = 10000 N × 1 m = 10000 J.

His average power output = (work done) ÷ (time taken) = 10000 J ÷ 200 s = 50 W

However, an electric winch might do the same job in 5 seconds, in which case the amount of work done would be the same, but the power expended by the motor would be greater, since it takes less time. In the case of the winch, power = 10000 J ÷ 5 s = 2000 W or 2 kW.

There are countless instances where increased power at our command has made our lives easier. More powerful cars get us where we want to go, faster. More powerful trucks get bigger loads delivered. More powerful digging machines get canals dug faster. More powerful woodworking machinery gets logs reduced to boards without the sweat and discomfort of sawing them by hand, and in much less time. The quicker we get results, the more outcomes we can achieve in a given time, and the more profit we can make. Mechanical power, and the quest for yet more power, is of intense importance in the world of engineering.

How much power can a human being exert?

In order to appreciate what is meant by amounts of power such as 30 watts, 2 kW or 8 MW, we need some benchmark of how much power a human being can expend.

Only about 250 years ago, in the days before the advent of combustion engines, most work was done either by humans or by horses or oxen. There were also waterwheels and windmills providing power for milling grain, but no-one had bothered to measure how much power they put out. It simply wasn't necessary. If there was sufficient power to do a job, that was good enough. And the power provided by wind and water was free, too.

When steam engines were developed, potential users of these engines needed to compare different engines according to their qualities. They needed some measure of how much power an engine could develop. And they needed to be convinced that a steam engine could provide more power than their relatively expensive teams of horses could.

James Watt, after making observations of draft horses in action, performed some calculations and estimated the amount of work that an average horse could do in a given time.

He decided that the average horse could put out 33000 ft-lbs/min in a sustained manner while working, and that figure became the standard which he called the 'horsepower' [HP]. A one-horsepower machine can put out roughly the same amount of power as one average horse, working in a steady, sustained fashion.

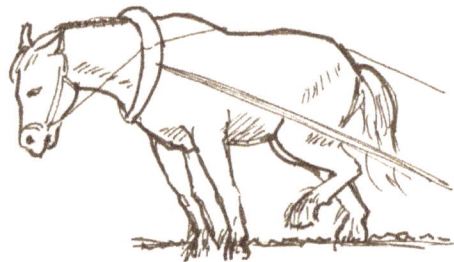

Since the introduction of S.I. units, power is now measured in watts [W]. It works out that one HP is equivalent to approximately 746 W. To get an idea of how much power this is, we need to compare it with the amount of power we ourselves can expend.

You can easily test your own power output. One way of doing this is to time yourself running up a flight of stairs, of known height.

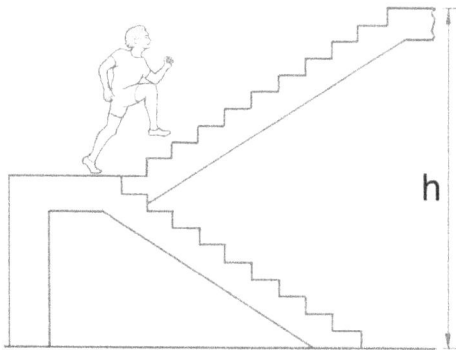

Suppose you weigh 800 N, and can run up several flights, with an altitude gain of 6 metres, in 10 seconds.

The mechanical work done is 800 N × 6 m = 4800 J. Your power output = (work done) ÷ (time taken)

= 4800 J ÷ 10 s = 480 W, or about ⅔ of one HP.

This doesn't sound too bad, being ⅔ as powerful as a horse. However, you could not sustain that rate of expenditure of energy for very long.

It would be quite a different matter if you had to go up 20 flights of stairs. Of necessity, you would have to go slower, and might take, say, an average of 30 seconds for every 6m of height gained. This would make your power output about 20% of one horsepower.

If you are trained and fit, you might expend energy in a short burst, say during a 100 m sprint, at a rate of approximately 2.5 HP, or 1500 W. Very few humans could keep this up for more than half a minute. If you needed to perform work at a sustained rate for hours on end, you might manage an output of only 50 W. Knowing this helps us to get a better conception of the magnitudes of the power ratings of motors and appliances.

For comparison, a 10 kW motor (equivalent to 13.4 HP) can accomplish the same amount of work in the same amount of time as could 67 men, working at a sustainable rate.

By a similar analysis, a 60 MW electricity-generating power station has the equivalent power output of 402 000 men.

Basic power calculation examples

Example 1

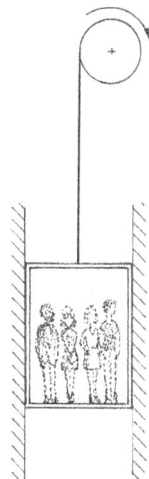

Determine the work done by a motorised winch in raising a lift cage of mass 800 kg with a payload of 400 kg, through a height of 40 m, at slow constant velocity. Assume the friction in the guides is 460 N. If this motion has to be accomplished in 25 seconds, what should be the power output of the motor?

The force required is (1200 × 9.81) + 460 = 12 232 N

Work done = force × distance = 12 232 × 40 = 489280 J

Power = (work done) ÷ (time taken) = 489280 ÷ 25 = 19571 W

= 19.57 kW. Say, 20 kW.

In practice, the motor chosen for this task would have to be a lot more powerful than this, because it would need to expend extra power when accelerating the lift up to speed. Also, the load carried by the lift might sometimes exceed the stated maximum load.

How many men acting together would be needed to accomplish this task in the same time? If each man could expend power at a rate of 150 W (approximately 0.2 HP) the number of men needed would be 19571 ÷ 150 = 131

Example 2

Determine the average power expended by a railway engine to accelerate a train of mass 600 tonnes up an incline of 1 in 100, uniformly from rest, to reach a velocity of 81 km/h, over a distance of 1200 m. The rolling resistance is 60 kN. Ignore air resistance. (Railway inclines, as distinct from gradients, are ratios of height difference to distance moved *along* the track.)

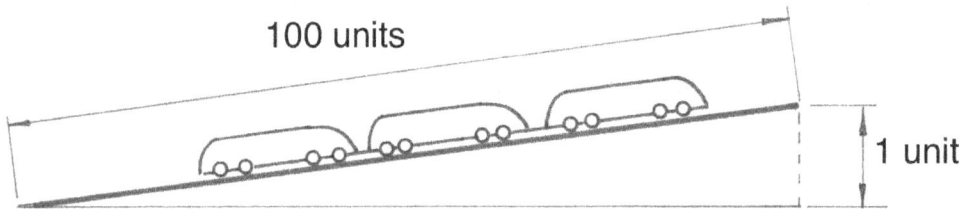

81 km/h = 22.5 m/s

Acceleration, a, is determined from
$v^2 = u^2 + 2\,a.s$ where in this case, u = 0
$\therefore a = 0.2109$ m/s^2

The time taken, t = (v - u)/a = 22.5 ÷ 0.2109375 = 106.7 seconds

The effective driving force in [N] = $mg.\sin\theta + ma + 60\,000$

= (600 000 × 9.81 × 1/100) + (600 000 × 0.2109) + 60 000 = 245 423 N

Work done = force × distance = 245423 N × 1200 m = 294.5 MJ

The average power expended by the engine = total work done ÷ total time taken

= 294 507 000 J ÷ 106.7 s = 2 761 003 W = 2.761 MW

Average power and maximum power

The preceding example leads us to look at the difference between the *average power* and the *maximum power* put out by the engine. The rate at which work is done changes with the speed of the train. Even though the driving force remains constant, when the train is going slowly, there is relatively less work being done in a period of given duration, than when it is going fast. So, the power output depends on the speed.

For any vehicle, moving at velocity 'v':

power = (work done) ÷ (time taken)

      = (force × distance) ÷ (time taken to cover that distance)

      = (force) × (distance ÷ time taken to cover that distance)

      = force × velocity

Hence, the instantaneous power output of any vehicle can be determined if the driving force, as well as the velocity at that given instant, is known.

Consider a vehicle that accelerates from rest up to a maximum velocity, after which it continues at this velocity.

When the vehicle reaches this maximum velocity, the driving force supplied by the engine drops suddenly. It doesn't disappear altogether, because some driving force is still needed to overcome air resistance and rolling resistance. Also, if the vehicle is going uphill, some driving force is needed to counteract the downhill weight component of the vehicle.

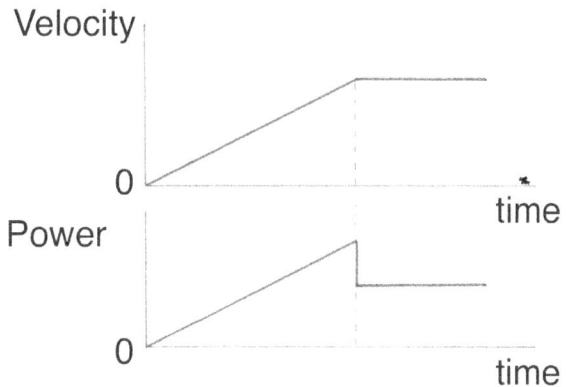

This graph shows that the maximum power expenditure occurs at the instant the vehicle reaches its maximum velocity.

For the train in the above example, the maximum power output will occur at the instant the train reaches its maximum velocity, and is given by

power = force × velocity = 245423 N × 22.5 m/s = 5 522 018 W = 5.522 MW

Exercises on basic power calculations

Question 1

A motor-car of mass 1000 kg experiences rolling resistance of 300 N, assumed constant at all speeds. Assume the air resistance on this vehicle is 200 N when it is moving at 72 km/h. Determine the power output required to drive this vehicle:

a. At a constant velocity of 72 km/h along a level road,   [10 kW] and
b. At 72 km/h on a level road, while accelerating at 2 m/s^2. [50 kW]

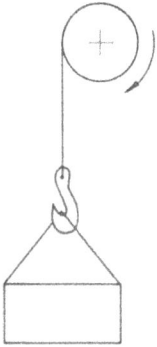

Question 2

Determine the power output required of an electric hoist, if it is to raise a load weighing 20 kN a height of 5 m in 10 seconds. [10 kW]

How many men would it require to perform the same task in the same time? Assume they would each be expending power at a rate of 0.5 HP for that short time.   [27 men]

Question 3

A carpenter uses a hand plane to square off some rough-cut timber. If he needs to push the plane with an average force of 120 N on each stroke, and he performs 320 strokes, each of 500 mm length, how much mechanical work does he do? [19200 J]. If this task takes 20 minutes to complete, what is his average power output?   [16 W]

Work Done by a Torque

Suppose a torque T acting on some rotatable object such as a shaft, is represented by a force P acting tangentially at a radius r from the axis of rotation. We define $T = Pr$.

If the torque turns the shaft through θ radians, then the point of application of the force P moves a distance s along the circumference of its circular path, where $s = r\theta$.

Work done = force × distance = $Ps = Pr\theta$

However, $Pr = T$ *by definition*

Therefore, the work done by a torque T in turning a shaft through θ radians:

Work done = $T\theta$

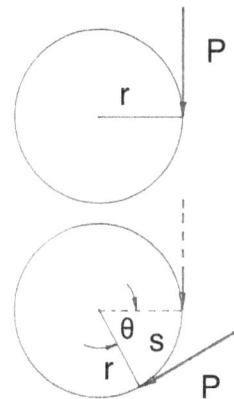

Example:

How much work is done by a torque of 25 Nm in turning a shaft through 100 revolutions?

100 revolutions = 100 × 2π radians

Work done = $T\theta$ = 25 N.m (100 × 2π rad) = 15708 J = 15.71 kJ

Power transmitted by a torque

Power is the rate at which work is done. We have seen that the amount of work done when a torque T turns a shaft through θ radians, is Tθ joules.

So, if the shaft turns at an angular velocity ω = θ radians/second, the amount of work done in one second is Tθ joules.

So the rate at which power is expended is (work done/time) = Tθ/t = T(θ/t) = Tω

Hence, the power transmitted by the torque turning a shaft at ω rad/s is:

P = Tω   namely, power = torque × angular velocity

Example: How much power is being expended when a torque of 15 Nm is turning a shaft at 160 r/min?   Power = Tω = 15 (160 × 2π/60) = 251.3 W

Exercises on power in systems where rotation occurs

Question 1

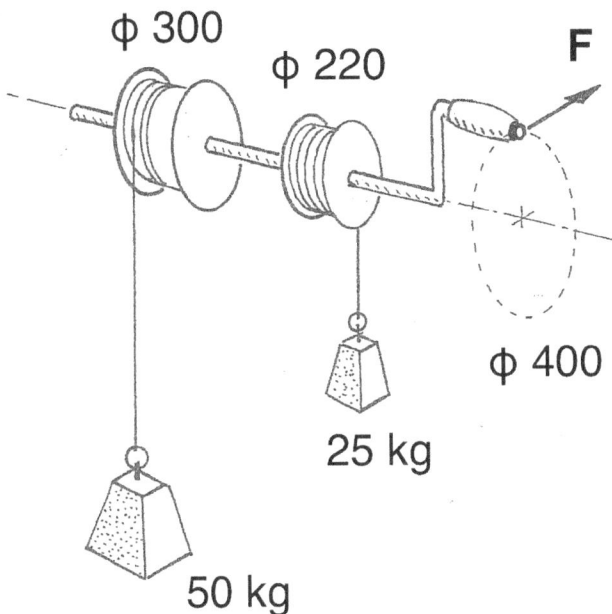

φ 300

φ 220

F

φ 400

25 kg

50 kg

Two winding drums are fixed to a shaft that has a crank at one end. The shaft is mounted in bearings (not shown) that have a frictional torque of 5 Nm altogether.

Exerting tangential force F, a man winds the crank to raise the 50 kg mass-piece a height of 4 m, at slow constant velocity.

Determine:

a. The magnitude of force F [207.9 N]

b. The work done by the man [1109 J] and

c. How long it would take, if he is capable of sustaining a power output of 0.1 HP. [approximately 15 seconds]

Question 2

Using a hand-cranked winch, a man pulls a trolley full of gravel, at slow constant velocity, up a ramp angled at 15° to the horizontal. The trolley weighs 230 N and its rolling resistance is 50 N. The frictional torque in the winch bearings is 16 Nm. The gravel weighs 880 N. the drum diameter is 220 mm. The crank handle radius is 380 mm.

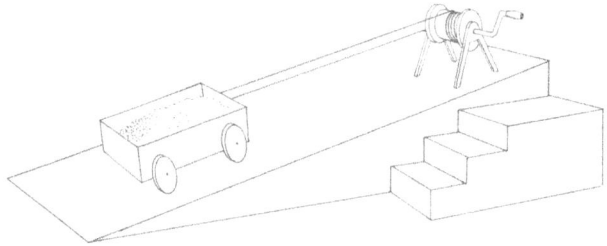

The wire rope used to pull the trolley remains parallel to the surface of the ramp. Ignoring the weight of this rope, determine:

a. The tension in the rope sufficient to cause the trolley to move. [337.3 N]
b. The magnitude of the tangential force he needs to exert on the crank to get the trolley to move. [140 N]
c. The work that must be done to raise the trolley a distance of 20 m along the ramp. [6746 J]
d. The time it will take the man to pull the trolley this distance, if he can sustain a power output of 100 W. [68 seconds]

Mixed exercises on work, energy and power

Question 1

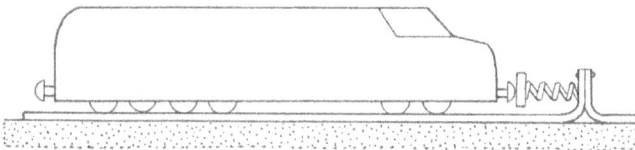

A railway engine of 125 tonnes, shunting in a siding, hits a buffer plate at the end of the track while freewheeling at 0.5 m/s.

The engine is brought to rest with uniform deceleration, and, in the process, compresses the spring behind the buffer plate by 112 mm.   Determine:

a. The maximum force exerted on the buffer plate. [139. 5 ×10^3 N]
b. The work done on the spring   [15. 625 kJ],
c. The stiffness of the spring   [2.491 MN/m],
d. The amount of compression in the spring required to bring to rest a 200 tonne engine moving at 1.2 m/s   [340 mm],
e. The magnitude of a steadily-applied force that would produce the same amount of compression in the spring as in d. above. [847 kN]

Question 2

A steel ball of mass 1 kg rolls from rest, down a hard-surfaced continuous slope, without slipping, from point A to point B, where it collides with a 2 kg ball, causing the second ball to roll down a further continuous slope BC, which is surfaced with rubber that offers some resistance to rolling.

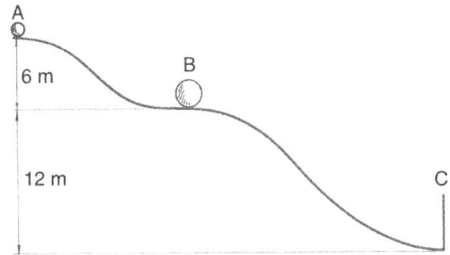

The 1 kg ball remains stationary after colliding.

Assuming no energy is lost in the collision, determine:

a. The velocity of the 1 kg ball immediately before the collision [10.85 m/s]

b. The amount of work that has been done against friction on slope BC, if the 2kg ball hits the wall at C with a velocity of 12.4 m/s. [140.5 J]

Question 3

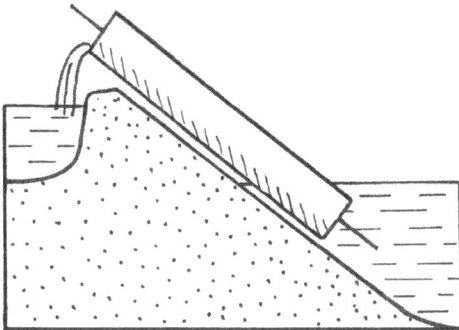

A motor drives an archimedes' screw, running in a tube, to raise water from a dam to an irrigation canal. The water emerges from the tube 3.2 m above the surface of the dam.

If the motor operates continuously with an output of 500 W, and the system is 35% efficient, how many litres of water can be delivered per minute?

Assume the velocity of the emerging water is negligible.

35% efficient means that only 35% of the energy input is turned into useful work done. [334.5 litres]

Question 4

A railway engine of mass 75 tonnes is allowed to freewheel down an incline of 1 in 60, and does so with slow constant velocity. Ignoring any effects of air resistance, determine:

a. The rolling resistance of this engine [12.26 kN]

b. How much work needs to be done to accelerate this engine uniformly, from rest, up an incline of 1 in 40, to reach a speed of 72 km/h over a distance of 2.2 km? [82.45 MJ]

c. What percentage of this work is done against inertia, gravity and rolling resistance, respectively? [18.1 %; 49.6 %; 32.7 %]

95

Question 5

A hand-cranked grindstone with diameter 600 mm is operated at steady constant velocity by one person turning the crank (radius 200 mm) while another person presses a tool to be sharpened against the grindstone.

The crank is connected to the axle of the stone by gearing: the larger gearwheel has 40 teeth and the smaller has 20.

600

200

The coefficient of friction between the tool and the grindstone surface is 0.6. The tool is pushed against the stone with a normal (that is, radial) force of 20 N. Assume there is no energy lost in the transmission of power from the crank handle to the stone.

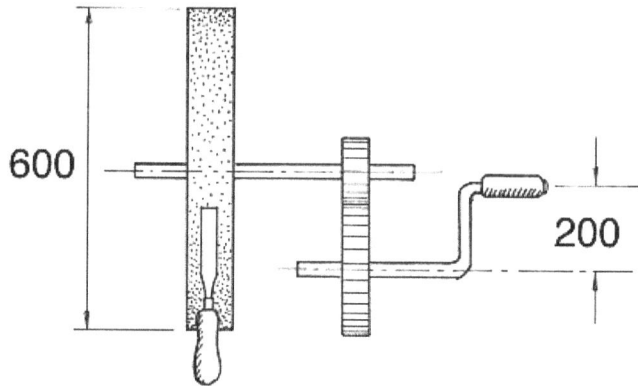

Determine:

a. The tangential *force* that the crank handle must be pushed by, to maintain constant velocity. [36 N]

b. The angular velocity of the crankshaft if the person operating the crank is maintaining a steady power output of 1/20 horsepower. [5.181 rad/s]

c. The tangential velocity of the crank handle under these conditions. [1.036 m/s]

d. The linear velocity of a point on the rim of the grindstone. [3.109 m/s]

e. The amount of work done on the tool edge, if the system is operated like this for five minutes. [11.19 kJ]

f. Describe what happens to the work done on the tool edge. Some of it goes into deforming the material of the tool. What happens to the rest of it?

Question 6

A compression coil spring has a natural (uncompressed) length of 200 mm, and a spring constant of 4068 N/m.

a. How much work must be done to compress this spring from a length of 160 mm to a length of 90 mm? [21.36 J]

b. If this spring were placed inside a tube, on a floor, in a vertical position, and compressed to the point where the top of the spring is 60 mm above the floor, how high above the floor could it shoot a ball with mass 200 grams? [20.32 m]

Energy considerations in

Gradual loading, sudden loading, and shock loading

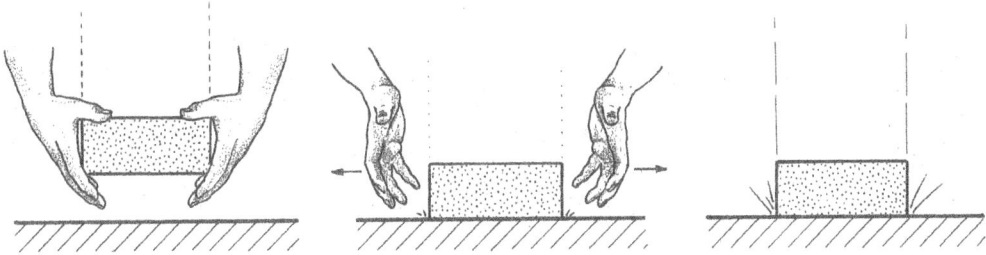

A given load descending from above can reach a support in one of three ways: it can be gradually lowered onto the support, or placed in contact and suddenly let go, or be dropped (or even projected) onto the support. The effective loading experienced by the support is different in all three cases, and depends upon how much mechanical energy possessed by the load is available to do work on the support.

What happens when you lower a load gradually

Suppose you take a package off a table and lower it slowly to the floor. Are you performing work?

During the controlled, slow downward motion of the package, the package has to be in equilibrium. Hence, for the duration of that movement, you have to be exerting an upward force, P, equal to the weight of the package.

However, instead of the force you are exerting advancing in its own direction, which happens when you *raise* a load, your effort is being pushed *back* along its own line of action. This means that gravity is, in fact, performing work *against* your effort.

When the package reaches the floor, it has less PE than before, as it now occupies a lower position in the Earth's gravitational field. However, in contrast with a situation in which the package were to *fall* from the table, a lowered package reaches the floor with *zero* gain in KE.

In the process of lowering, you have expended energy, but you have not added to the mechanical energy of the package. This energy that you expended effectively *neutralised* the gain in KE that would have occurred if the package had descended to the floor without your intervention. The energy that you put in cannot be retrieved.

So, an object that is lowered gradually has its available mechanical energy diminished by an amount equivalent to the PE that it gave up by moving to the lower altitude.

In fact, the only energy that a lowered object possesses is whatever PE remains to it in its new position. The only force it can exert on the support is equal to its own weight.

Gradual vs. sudden loading

If you place a load only just in contact with the support, and let go of it suddenly, the effective loading on the support is *greater* than that caused by a gradually lowered load. We can see how much greater this loading is, by examining what happens to a coil spring, when a mass-piece is placed upon it, respectively gradually and suddenly.

Firstly, gradual loading:

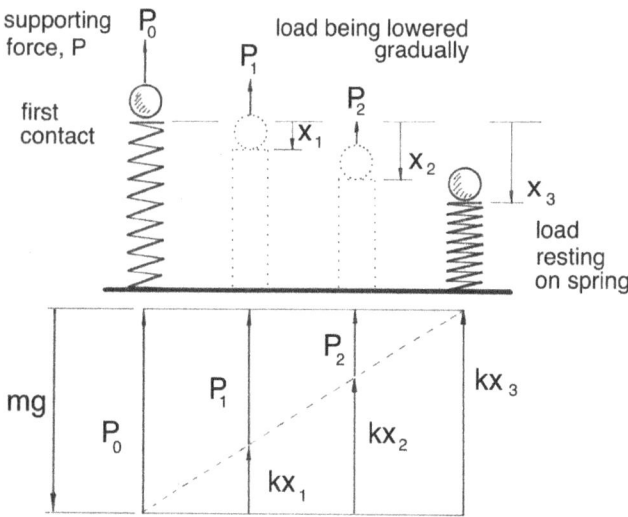

Suppose a mass 'm' is lowered gradually onto a compression coil spring of stiffness 'k'.

At the start of the lowering process, you have to exert an upward force P_0, needed to balance the weight of the mass-piece.

As you lower the mass, the spring becomes gradually compressed.

The force in the spring gradually increases until the spring is supporting the entire weight of the mass-piece.

Suppose the amount of compression in the spring at that time is 'x'. (Shown as x_3 above)

During the time that the force exerted by the spring increased from 0 to the value kx, the supporting force that you needed to exert decreased linearly from P = mg to P = 0, as the spring took over the support from you. By exerting this diminishing force P, over the distance 'x', you have neutralised the gain in kinetic energy that the load would have experienced by descending this distance 'x'.

So, the only downward force that the load can exert on the spring is that due to its own weight, mg. This is balanced by the compressive force in the spring, so we have: mg = kx, or, looking at the amount of compression that resulted: x = mg/k

In the case of a sudden loading:

The spring takes the full weight of the mass-piece immediately you let it go. There is no diminishing upward force P to partially support the load. So, *all* of the available PE of the load goes into compressing the spring, and becomes stored in the form of elastic energy.

The amount of PE given up would be mgx, so: $mgx = \frac{1}{2}kx^2$ $\therefore x = 2\,mg/k$

This means that the compression in the spring produced by sudden loading is exactly twice that produced by gradual loading. And it follows that the force produced in the spring will be exactly twice the force that would have resulted from a gradual loading.

Shock Loading

If you *drop* a ball onto a spring from a height 'h', the amount of potential energy available to be stored as elastic energy is given by:

$PE = mgh + mgx$, where x is the maximum compression resulting from this drop.

Assuming for the moment that no energy is lost in the impact when the ball hits the spring (a rough approximation, since *some* energy is *always* lost in an impact): if no energy is lost, then all the available PE would be converted to elastic energy,

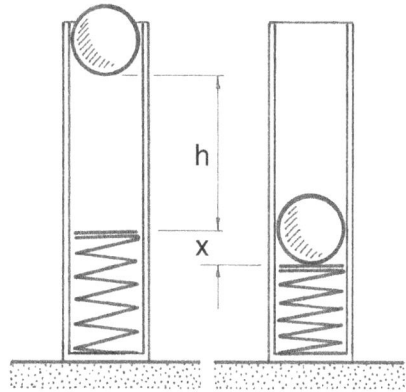

$\therefore mgh + mgx = \frac{1}{2}kx^2$ yielding the quadratic equation: $(\frac{1}{2}k)x^2 - (mg)x - mgh = 0$

which can be solved for x, provided that the values of k, h and m are known.

Example

A steel ball of mass 1 kg is dropped from a height of 600 mm onto a coil spring of stiffness 1800 N/m, which is contained in a vertical tube with a close fit. Determine:

*   The maximum amount of compression in the spring before it rebounds,
*   The maximum force in the spring, and
*   The number of times this force is greater than the static weight of the ball.

Assume firstly that no energy is lost in the impact, and secondly, that 30% of the available PE is lost in the impact.

Firstly, assuming that all the PE is converted to EE: $mgh + mgx = \frac{1}{2}kx^2$

$\therefore (\frac{1}{2}k)x^2 - (mg)x - mgh = 0$

$\therefore \frac{1}{2}(1800)\,x^2 - 1(9.81)\,x - 1(9.81)\,(0.6) = 0$

$\therefore 900\,x^2 - 9.81\,x - 5.886 = 0$ $\therefore x = 0.08650\ m\ = 86.50\ mm$

Max force F = kx = 1800(0.08650) = 155.7 N

Number of times this force is greater than the static weight of the ball:

= 155.7 N ÷ 9.81 N = 15.87 times

The reader can confirm that the equivalent answers for a 30% loss in energy are:

71.58 mm; 128.9 N; and 13.13 times the static weight of the ball.

Exercise

A steel ball of mass 2 kg is dropped from a height of 1200 mm onto a coil spring of stiffness 5000 N/m, which is contained loosely in a tube. It is observed that the maximum compression in the spring before it rebounds is 80 mm.

Determine:

The percentage of the potential energy of the ball that is lost in the impact with the spring. [36.3 %]

The maximum force in the spring. [400 N]

How many times greater is this force than the static weight of the ball? [20.39 times greater]

How far from the top of the tube the ball would rise on its rebound. [464.5 mm]

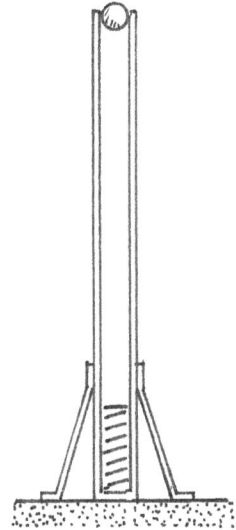

Exercise

A hard rubber ball of mass 5kg is dropped a height of 2.0 m onto a hard floor from rest at point A, and rebounds to a height of 1.4 m before falling back.

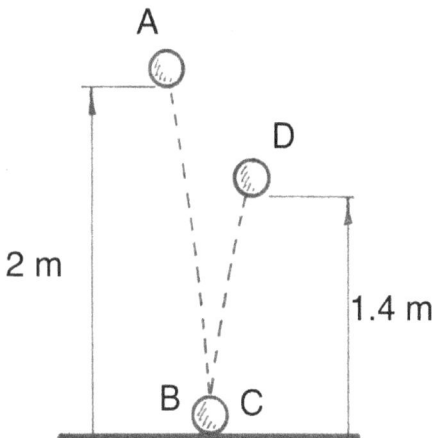

Determine:

- The velocity of the ball in position B, just before impact with the floor. [6.264 m/s]

- Its velocity in position C, immediately after leaving the ground. [5.241 m/s]

- The amount of energy lost by the ball, on impact with the floor. [29.43 J]

- The height of the next bounce, if it loses the same fraction of its total energy every time it hits the ground. [0.98 m]

Tutorial questions on energy applied to shock loading

suitable for solving in small groups

Question 1

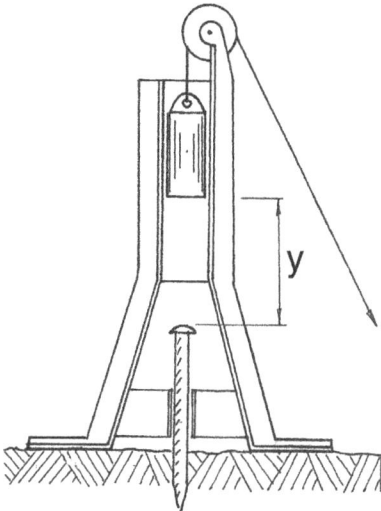

A pile of negligible mass must be driven into the ground. A pile driver is set up, consisting of a solid steel cylinder of mass 100 kg, which is raised by four men pulling on a rope passing over a sheave at the top of a tower.

To operate the pile driver, the men release the rope. The cylinder descends between guides which ensure that it transmits a vertical force to the pile.

The dimension 'y' is 4.00 m at the start of the first drop. It takes 20 drops to drive the pile a distance of 1.00 m into the ground.

Assume that no energy is lost in friction at the sheave, and that the men are capable of working at a sustained energy output of 50 W.

Also assume that between drops of the pile driver, it takes the men 5 seconds to get hold of the rope again and start pulling. Determine:

a. The average resisting force of the ground.   [89.28 kN]

b. The average time taken to raise the cylinder between drops. [22.3 seconds]

c. How long the entire process is likely to take. [between 9 and 10 minutes]

Question 2

A brick of mass 3 kg is dropped a height of 5 m onto levelled ground. The brick lands squarely on one face, and penetrates to a depth of 10 mm before coming to rest. Assuming that it experiences uniform deceleration after its first contact with the ground, and does not break upon impact, determine:

a. The velocity before impact. [9.905 m/s]

b. The magnitude of the deceleration. [4905 m/s^2]

c. The average resisting force exerted by the ground.   [14.72 kN]

d. The amount of energy given up by the brick during its fall. [147.4 J]

e. The mass of a stack of bricks with the same 'footprint' that would produce the same size depression in the ground if placed down gently instead of being dropped. Hint: the loss in PE of the stack of bricks must equal the energy required to deform the ground in the same way. [1503 kg, representing 501 bricks of this type]

Chapter 14

Simple Lifting Machines (SLMs)

The definition of SLMs and why engineers need to understand them
Mechanical advantage and velocity ratio
Using an energy accounting diagram to illustrate the efficiency of a SLM
The 'law' of a machine
Specific characteristics of several well-known types of SLM
Exercises and group work tutorials
Lab experiments and design-and-build projects dealing with SLMs

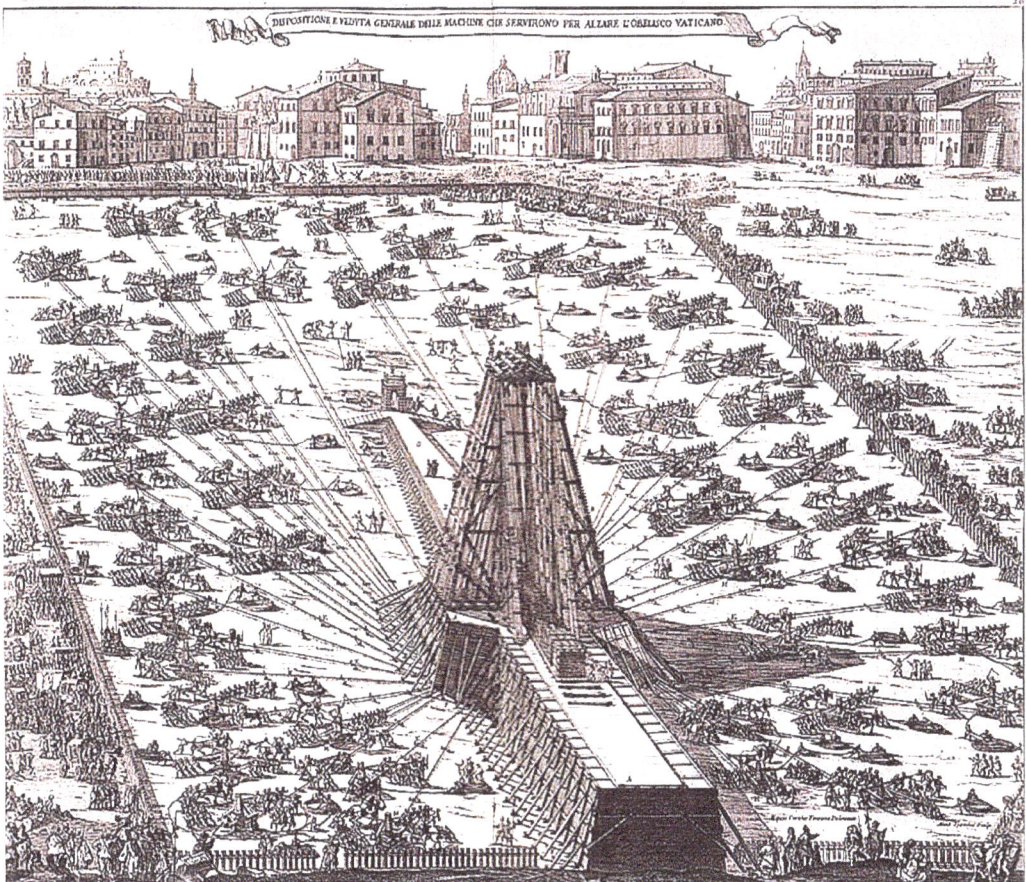

DISPOSITIONE E VEDUTA GENERALE DELLE MACHINE CHE SERVIRONO PER ALZARE L'OBELISCO VATICANO

A contemporary illustration showing the raising of an obelisk in Rome in 1586. All the visible ropes are parallel with the ground, so how do you think this process worked?

The definition and purpose of simple lifting machines

Quite early in human history, people saw a purpose in raising heavy loads, like felled trees and large stones used for building: loads too heavy for one man to handle alone. Where possible, teams of men would be deployed to work together in raising the loads. The source of power was almost always muscular contraction exercised by humans or animals.

Over time, devices were developed to ease this task. Such devices employed basic mechanical components that have been in use for millennia, such as wedges, levers, inclined planes, rollers, sheaves and ropes. A great deal was accomplished using such elementary components.

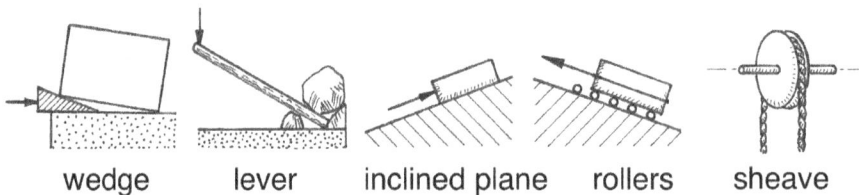

| wedge | lever | inclined plane | rollers | sheave |

The word 'sheave' may be new to some readers. A sheave is a wheel-like disc that is free to turn on its axle, and has a groove in its rim. The groove is for a rope, string or belt to pass around the sheave, without slipping off it.

There is no difference in shape between a pulley and a sheave. Many people use these two words interchangeably. This author prefers to call such an object a 'pulley' when it is used to transmit power in a machine in continuous rotary motion, and a 'sheave' when it passively re-directs a rope passing over it. However, usage differs all over the world, so call a sheave a pulley if you like. Everyone will know what you mean.

Most lifting machines in modern industry are 'complex' machines, in that they are made up of many parts, and are powered by electric motors, hydraulic systems, or combustion engines. A 'complex' machine frequently also incorporates a computerised control system. Machines like this are used for lifting large loads, such as: moving building materials, raising ore from mines, handling heavy items in factories, loading and unloading shipping containers, and jacking up vehicles.

Simple lifting machines: In contrast to a complex lifting machine, a 'simple' lifting machine is entirely mechanical, has a relatively small number of parts, and uses human muscular contraction as a source of force. SLMs are used in applications where the loads to be raised are relatively small.

Examples of simple lifting machines still in widespread use today include winches, screw jacks and block and tackle systems.

Any simple lifting machine is made up of a combination of the age-old basic mechanical components mentioned previously, sometimes including relatively

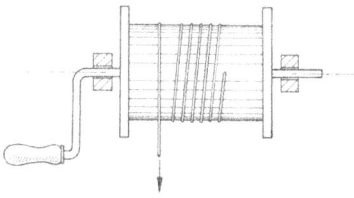

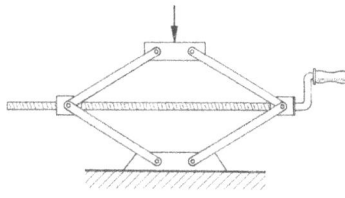

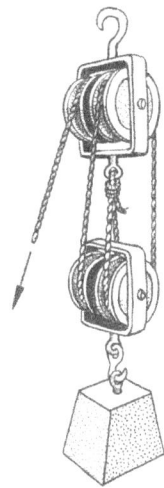

newer devices such as crank handles, gear teeth and screw threads.

Simple lifting machines can be constructed using many different arrangements of these basic components.

Some SLMs, such as motor car jacks, are permanent assemblies of parts, kept for repeated use. Others, such as block and tackle assemblies, can be temporary arrangements, put together for a specific purpose and dismantled afterwards.

The importance of studying simple lifting machines

Three reasons why it is essential for engineering students to study the working of simple lifting machines:

- The working of a SLM provides a practical example of the application of the principle of conservation of energy. By analysing the operation of a simple lifting machine, we get to understand the basic principles of energy flow that apply to *all* machines, including complex machines

- With some hands-on experience, we can learn to assemble basic machine components to construct a simple lifting machine that can be put to practical use. At the very least, we should be enabled to understand those SLMs that we encounter in industrial applications.

- The procedure that we use to analyse the variables involved in the operation of a SLM gives us practice in thinking 'like an engineer'.

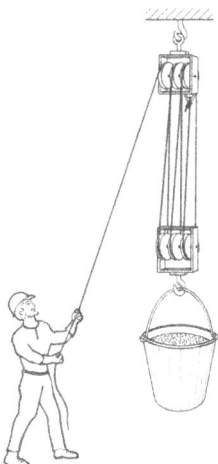

The mechanical advantage (MA) of a simple lifting machine

The muscular force a person exerts in order to operate a SLM, is called the *effort*. We only need a lifting machine if the load we need to raise is *greater* than the force we can exert. Therefore, it should be obvious that the effort applied to any lifting machine would always be *smaller* than the load.

If a machine makes it possible for a small effort force to raise a large load, we regard this machine as effective. The measure of this effectiveness is the mechanical advantage,

MA, which is defined as

MA = (load) / (effort) = L ÷ E

For example, suppose a man using a block and tackle has to exert a force of 220 N in order to raise a load of 1210 N. The mechanical advantage of this machine would be:

MA = load / effort = 1210 N / 220 N = 5.5

Clearly, it is desirable for a lifting machine to have a high mechanical advantage, if possible. Simple lifting machines usually have values of MA ranging from slightly greater than 1, to, at the very most, around 400. While it is quite possible to design a SLM with values of MA exceeding 400, such a machine would not be not practical, for reasons that will become clear later.

Work input and output on a SLM

A simple lifting machine can give you a bigger output *force* than the input force, but it cannot magically multiply the energy that you expended while operating the machine. The mechanical work done by an effort force is an input of energy. Since a simple lifting machine is subject to the law of conservation of energy, you can't get more work out of a SLM than you put into it.

In fact, due to inevitable energy losses in the operation of a SLM, you can only get slightly less work out of a machine than you put into it.

Suppose a simple lifting machine is used to raise a load of 800 N through a height of 2.5 m. The work done on the load to increase its PE would be 800 N x 2.5 m, namely 2000 J. At the very least, assuming no energy losses, we would have to put in *this same amount of work* to operate the machine. In practice, however, there are *always* energy losses, so we will actually have to put in *more* work than 2000 J.

Before we can obtain a perspective on exactly *how much* additional work will be needed to operate any lifting machine, we need to define a few more concepts related to SLMs.

The velocity ratio (VR) of a simple lifting machine

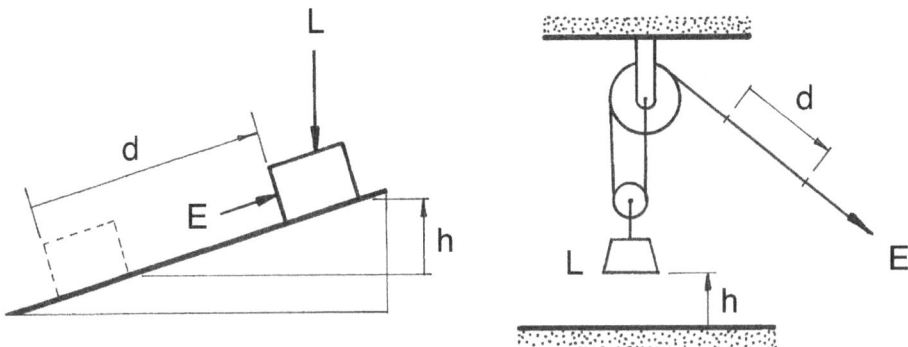

For any given SLM, the velocity ratio is defined as (distance moved by the point of application of the effort) divided by (vertical distance moved by load). **VR = d ÷ h**

It is important to note that the 'load distance' that we refer to is always the *vertical height through which the load is raised*. This is so because the very aim of a lifting machine is to raise a load against gravity. If we succeeded only in moving the load sideways, we would not have accomplished our aim. This distinction is vital in instances where the load moves both upwards, and sideways, such as on a inclined plane. The distance moved by the point of application of the effort does not have to be vertical. It can be in any direction, and can even be a circumferential distance, in cases where the effort is applied through turning a crank handle.

Although the velocity ratio is the ratio between two distances, it is called a 'velocity ratio', rather than a 'distance ratio', because we are speaking about the respective distances moved by the effort and the load *in a given time*. Clearly, the load must travel the load distance in *the same* amount of time that the effort takes to travel the effort distance.

The expression 'velocity ratio' (with the same essential meaning as for SLMs) is particularly appropriate when applied to rotating machinery, such as transmission systems. With the type of operation found in continuously rotating machines, the rotational velocities of the input shaft and output shaft can be more easily compared than can the distances moved by any specifically identified points.

In a SLM, the distance 'd' moved by the effort is always greater than the vertical distance 'h' moved by the load. You probably know this from various examples of lifting machines that you might have used.

For instance: 1. When operating a scissors jack to raise a car to change a wheel, the point of application of the force on the crank-handle moves many times the distance that the side of the car is raised. 2. When raising a load with a block and tackle, the effort rope has to be pulled down by a much greater distance than that by which the load is raised.

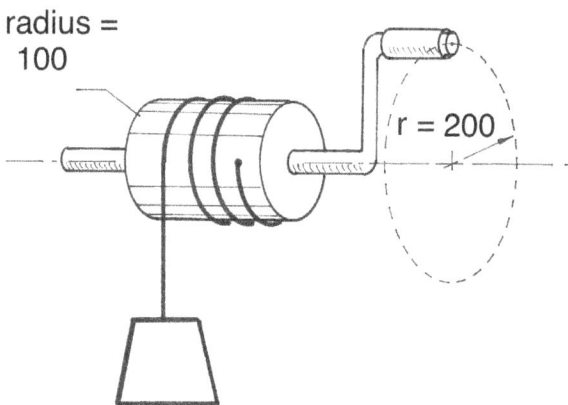

radius =
100

r = 200

3. Consider this simple hand-cranked hoist: to raise the load by a distance 'h', equal to one circumference of the drum, the effort force has to move a distance 'd', equal to one circumference of the crank handle's circle of movement. With the dimensions shown, the VR would be 2.

In any event, for a machine like this, or any other simple lifting machine, 'd' will always be greater than 'h'.

What if there were no energy losses in the operation of a lifting machine?

If a machine had zero energy losses, the output work would be exactly equal to the input work. Thus, we would have:

load × load height raised = effort × effort distance

L.h = E.d   ∴ L ÷ E = d ÷ h   ∴ MA = VR

Hence, for an ideal machine, with no energy losses, the mechanical advantage would be equal to the velocity ratio.

However, there is no such thing as a machine that loses no energy. In the operation of every lifting machine, there is the possibility of losing energy in the following ways:

- By having to overcome friction within the machine,

- By having to raise parts of the machine (such as hooks, slings and lower pulley blocks) that unavoidably have to be moved upwards in order to raise the load,

- By work done in stretching the ropes that take the load, where ropes are present, and

- By generating heat and sound (like making poorly-lubricated sheaves squeal)

The amount of energy lost due to heat and sound from the operation of a manual machine is usually negligible. However, the other three energy losses in the above list are usually significant, and need to be taken into account when we look at the energy flow during one operation of a lifting machine.

The energy accounting diagram

We can describe the energy input and output for one operation of a SLM, by means of an energy accounting diagram, which looks as follows, and is best drawn to scale:

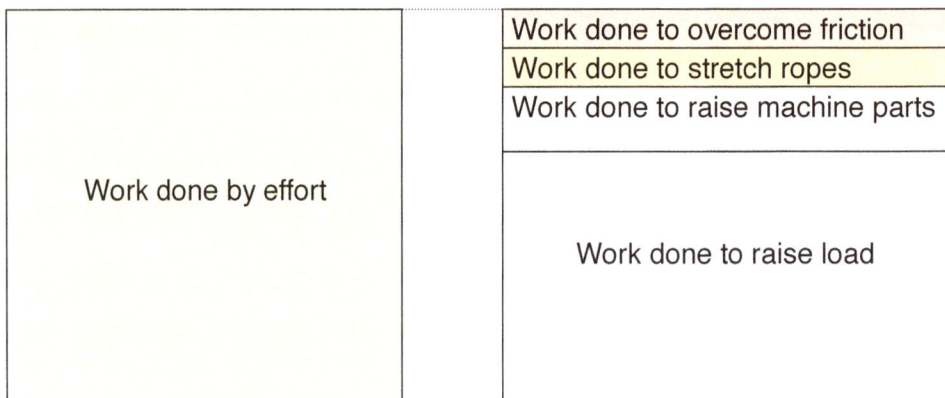

Work done by effort	Work done to overcome friction
	Work done to stretch ropes
	Work done to raise machine parts
	Work done to raise load

The height of the column on the left represents the amount of work done by the effort. What happens to all this work? By the principle of conservation of energy, the total work output has to be the same as the total work input. Therefore, the overall height of the column on the right must equal the height of the column on the left.

The concept of 'useful' work output from a SLM

The only directly *useful* work output of a lifting machine is the work done on the load to raise it. In the diagram above, the other three portions of the column on the right represent energy *losses*.

Strictly speaking, not all of the energy in these three 'loss' portions of the column on the right is irretrievably 'lost'. Under certain circumstances, a machine part that has been raised has the potential to do work by descending again, so that amount of energy can be recovered.

However, *none* of the work done to overcome friction can be regained, and it is extremely unlikely that work done to stretch ropes can be regained, so these two portions of the column represent permanent losses.

The energy accounting diagram, besides illustrating the way that input work is divided up into outcomes, will prove to be a useful tool for solving certain problems further on in this chapter.

How to determine the velocity ratio of a machine

The value of the velocity ratio of a given machine is determined by the dimensions and arrangement of the parts of the machine, and *this value is always the same for a given machine*.

(With the exception of machines that use ropes that stretch, because, where stretching occurs, the point of application of the effort will move further than expected for a given operation.)

To determine the value of the VR for any given lifting machine, one has to work out how far the point of application of the effort will move, compared to the vertical height that the load is raised in the same time. Once you know how each component of the machine affects the overall VR, you can determine the VR of the whole machine by calculation.

We start by looking at the velocity ratios of individual machine components:

a. A 'standing' sheave: one whose position is fixed relative to the frame of the machine.

It is obvious from the following diagram that if the load is raised by one unit of length, the point of application of the effort must move downwards by one unit. (Ignoring any stretch in the rope.)

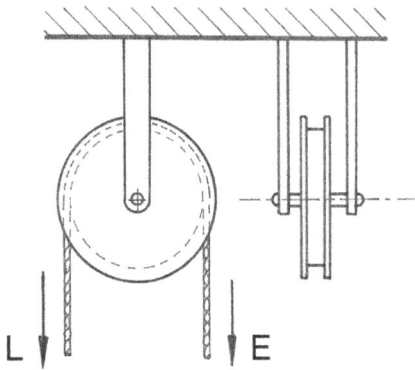

VR = effort distance/load height raised

= (1 unit)/(1 unit) = 1

Since the VR of a standing sheave has the value of 1, a standing sheave that is part of a lifting machine does *not* contribute to raising the VR of the whole machine.

Standing sheaves are included in lifting machines purely because they are needed to change the direction of the applied effort. This is their only function. It is easier for a human operator to pull downwards, than to pull upwards, so a standing sheave attached to a ceiling or to the frame of a machine is often a convenience.

b. A running sheave: one that moves upwards during the raising of a load.

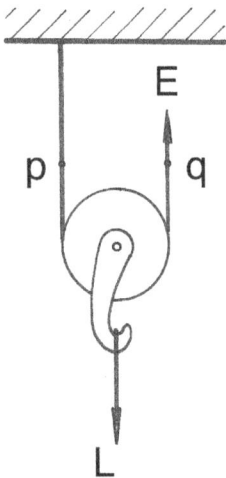

Consider two points on the rope: 'p' and 'q'. If the effort moves up by one unit, then point 'q' moves upwards by one unit, while point 'p' remains where it was. Hence, the loop of rope in which the sheave rests is shortened by one unit.

If you shorten a loop by any amount, then each side of the loop takes up half of the shortened amount. This means the sheave will rise by only half of one unit. The VR = effort distance/ load height raised = $1 \div (\frac{1}{2}) = 2$

To confirm this reasoning, it is a simple matter to set up a small running sheave with a light load attached (just to keep the sheave from tumbling over sideways), and to measure the distances moved respectively by the effort and the load.

For example, raise the effort end of the string until the load has risen by 200 mm, and see how far the point of application of the effort has moved. It ought to be 400 mm. If this distance turns out to be not exactly 400 mm, that could be due to the string having stretched, or inaccurate measurement of one or both distances.

Since the VR of a running sheave has a value of 2, every time a running sheave is incorporated into an assembly of machine components, it contributes to *increasing* the VR of the assembly,

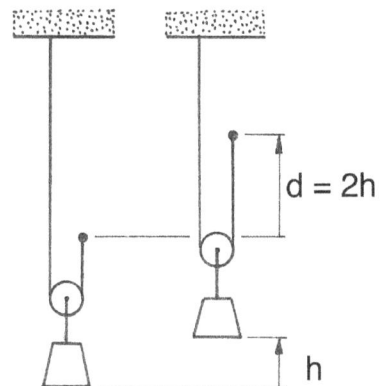

either by *adding* the value 2 to the overall VR, or by *multiplying* the overall VR by 2, depending on the way the sheaves are arranged. The difference will become clear when we examine different arrangements of sheaves in block-and-tackle systems.

c. A compound, or differential sheave

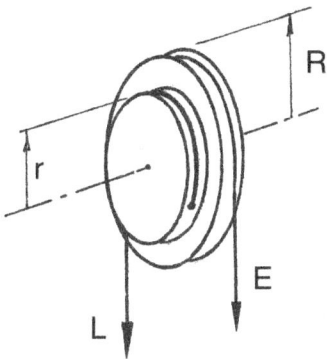

Consists of two co-axial sheaves (or drums) of different diameters, fixed together, turning on the same axle.

A rope is fixed separately to each sheave (or drum), and the two ropes are wound in opposite directions. The effort rope is attached to the larger drum, radius R, and the load rope is attached to the smaller drum, radius r.

If the differential sheave is given one full turn (one revolution), then the load is raised one circumference of the smaller sheave, namely $2\pi r$. In the same time, the effort rope is pulled downwards a distance equal to one circumference of the larger sheave, namely $2\pi R$. The VR = effort distance/load height raised = $2\pi R/2\pi r = R/r$.

The influence of sheave diameters

In all applications of sheaves in SLMs, it is important to note that the diameter of any single sheave *makes no difference* to the contribution that that sheave makes to the overall VR. The diameter is significant only when two sheaves are used as a compound sheave, when their relative diameters directly influence the VR.

In designing sheaves for particular applications, apart from the strength considerations, the following tendencies should be borne in mind: The smaller the sheave, the more acute the bend in the rope passing over it, and hence the greater the friction within the rope fibres that results from bending and straightening the rope, as it passes over the sheave. This particular cause of energy loss is more noticeable when stiff ropes are used. On the other hand, the larger the sheave, the more material has to be used to make it, and so, if use is made of running sheaves, the greater is the mass of the machine parts that will need to be raised. Also, large sheaves take up more room than small ones, thereby limiting the range of operation of the machine.

d. An inclined plane, at angle θ to the horizontal:

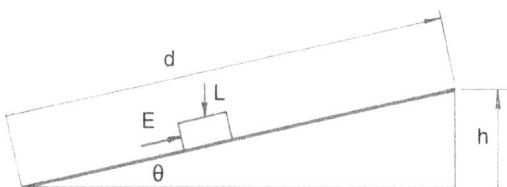

The effort moves distance 'd' in the time that the load is raised a distance 'h'.

VR = d/h, but h/d = sin θ and so the VR = 1/ sin θ.

e. Two meshing gearwheels.

The effort is applied in the form of a torque exerted by the driving wheel, and the load torque is provided by the resistance offered by the driven wheel.

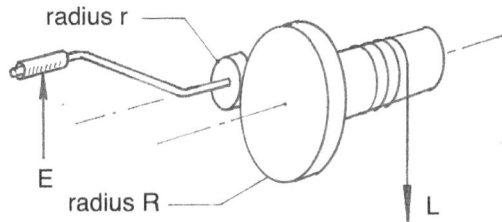

Two meshing gears on their own do not constitute a lifting machine, because we also need to know the radii at which the effort and load were applied, in order to determine by what distance the effort needs to move, for a given height that the load was raised. As shown here, these radii are usually different to the respective pitch circle radii of the two gears.

However, when pairs of meshing gears occur in lifting machines, they have an effect on the calculation of the VR, as follows:

In a given time, the larger wheel makes fewer turns than the smaller wheel, in inverse proportion to their radii. So, if the smaller wheel makes 'n' turns, the larger one makes (r/R)n turns. We will use this fact when we determine the velocity ratio of a given lifting machine containing gears.

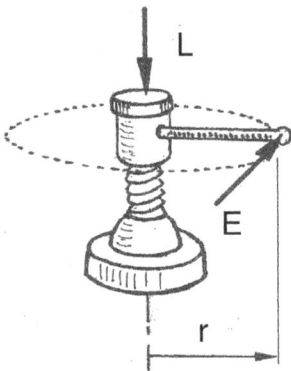

f. A nut on a screw thread.

By turning a nut on a vertical threaded shaft, we are able to make the nut progress upwards. If the nut supports a load, this will force the load to rise.

The effort force is applied tangentially, at a radius 'r', by means of either a rope around a pulley, or a rotating arm.

For one turn of the nut, the load rises an amount equal to the pitch, 'p', of the screw thread.

In the same time, the effort force moves through one circumference of its travel, namely a distance of $2\pi r$. The VR of this arrangement is thus $d \div h = (2\pi r)/p$.

How to determine the velocity ratio of any simple lifting machine made of several components.

1. Consider *one* operation of that machine. Start by supposing that during this operation, the load rises by a given amount, for example, by one metre. Alternatively, one can suppose that the shaft around which the load rope is wound, makes one revolution.

2. Determine, step by step, by how much each successive part of the machine will move during that operation.

3. This results in an expression for the distance that the point of application of the effort must move in the same time.

4. Compute the overall VR, from VR = (distance moved by effort)/(load height raised)

Example: determine the VR of this geared hoist driven by a hand crank.

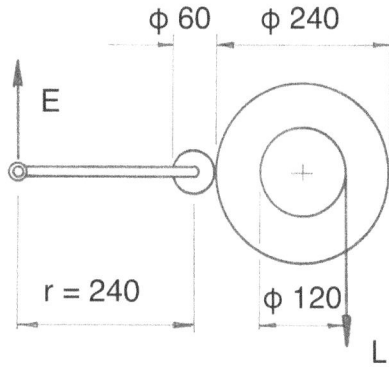

φ 60   φ 240

E

r = 240

φ 120

L

Let the load drum make one revolution. Then the load will rise by an amount equal to one circumference of the load drum, namely π(120) mm.

The large gear-wheel is fixed to the load drum, therefore also makes one revolution in this time.

The small gear-wheel will turn 240/60 times the number of revs turned by the large gear-wheel. This works out to 4 revolutions. The crank is fixed to the small gear-wheel, therefore also turns 4 revolutions in the same time.

The point of application of the effort force, E, moves 4 turns, times the circumference of its travel, namely 4 × π (240) mm.

VR = (distance moved by E) /(height that load L is raised) = (4π (240)) ÷ (π (120)) = 8

The velocity ratios of different arrangements of ropes and sheaves: The block and tackle

Arrangements known as 'block and tackle' are frequently used to raise loads or to tension ropes, particularly on sailing boats.

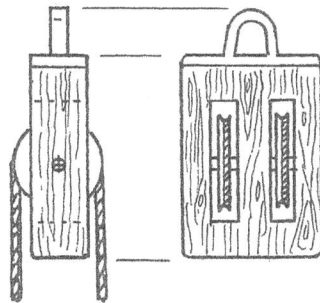

The word 'block' in this context refers to the frame in which the sheaves are housed. It is called a 'block' because, originally, these frames were carved from a solid block of hardwood, because that would resist the effects of sea-salt corrosion better than would the metals available at the time.

Today 'pulley blocks' are made of corrosion-resistant metal or advanced materials. They are no longer carved from a single block, but are made of several assembled parts, which can include ball or roller bearings. The sides of the frame of a modern 'block' are known as 'cheek-plates'.

The rope that is threaded through the blocks is called the 'tackle'. A lifting machine, or a device to tension a rope, can be made up of a number of different arrangements of sheave blocks and tackle. There are three traditional types of arrangement for blocks and tackle, called respectively a 'first system', a 'second system' and a 'third system'.

A 'First system'

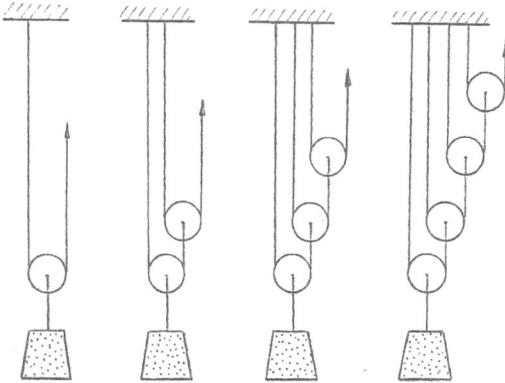

This is a system consisting entirely of running sheaves. In this arrangement, each running sheave *multiplies* the cumulative velocity ratio by a factor of two.

So, for the arrangements shown here, the velocity ratios are respectively 2; 4; 8 and 16.

A 'first system' that consists of more than two sheaves is too cumbersome to be used as a lifting machine.

There are too many parts that need to be lined up and mounted overhead. Also, each threaded sheave needs to be held in position while the next one is threaded. This makes it awkward to assemble. A 'first system' is best used as a rope tensioner, usually upside down compared with the arrangement illustrated above.

When used as shown here in a maritime setting, it has the additional advantage that the rope to be tensioned does not need to stretch by much, so the effort rope does not have to move an unwieldy long distance before it is secured.

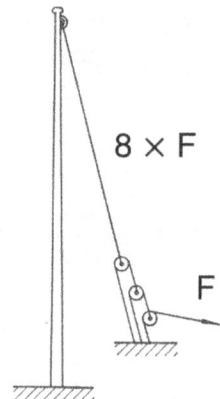

A 'Second system'

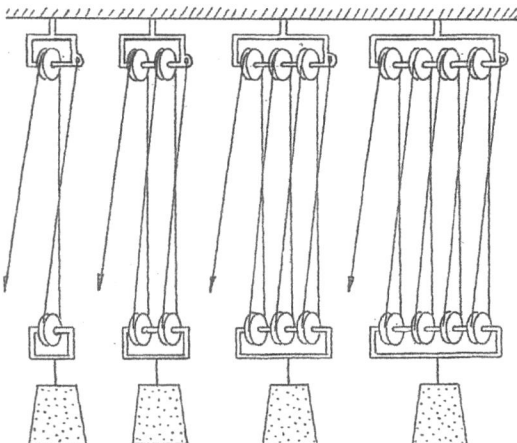

This is the most common arrangement for a block and tackle used for raising loads.

In this system each running sheave adds 2 to the cumulative velocity ratio.

So, for the arrangements shown here, the velocity ratios are respectively 2; 4; 6 and 8.

For a 'second system' like the arrangements illustrated here, a quick way of establishing the VR

114

is to count the number of ropes, not counting the effort rope. If the lower block is hanging from 2 ropes, the VR = 2. If it is hanging from 4 ropes, the VR = 4, and so on.

Another way of reasoning to remember this pattern is as follows: suppose there are 6 ropes supporting the lower block. If the load, and therefore the lower block, is raised by one metre, there will be 6 metres of slack rope that has to go somewhere. This amount of rope is taken up by the operator pulling downwards on the effort rope. So, VR = effort distance/load height raised = (6 m)/(1 m) = 6.

A variation on a 'second system'

You can produce a VR of 3 (or 5 or 7) by arranging to have only that number of load ropes.

Suppose you need a VR of 5. You would use a top block with 3 sheaves and a lower block with only 2 sheaves.

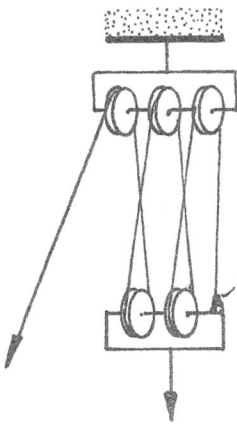

Instead of the rope end being fastened to the top block, as illustrated previously, it is now fastened to the lower block.

As before, the number of ropes supporting the load block is equal to the value of the velocity ratio.

A different way of arranging sheaves in a second system, to reduce friction.

When using a 'second system', pulling down on the effort rope tends to tilt the top block, with the result that the rope can ride out of the groove in the sheave and rub against the inside edge of the cheek-plate, causing extra friction.

One way to overcome this problem is to arrange all the sheaves so that they are in the same vertical plane, as shown below. With this arrangement, pulling on the effort rope does not tilt the system to one side.

friction occurs here

Note that this arrangement is mechanically identical to a 'second system' with two sheaves in the top block and two in the bottom block.

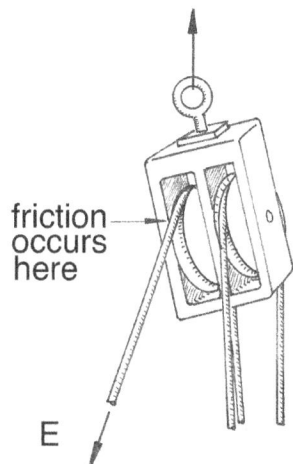

E

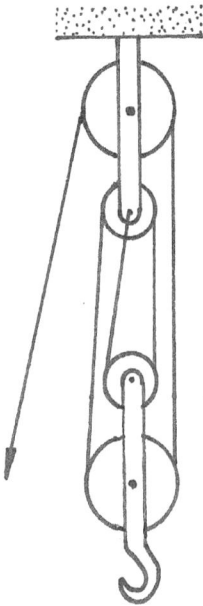

If one uses an all-in-line system like this, with three standing sheaves and three running sheaves, and the rope end is fastened to the top 'block', what do you think the VR will be?

Can you name a potential drawback of such an arrangement?

One practical limitation of a 'second system' of tackle

Very seldom does one find in use a pulley block containing more than 4 sheaves. The more sheaves there are, the more friction has to be overcome.

A block-and-tackle that has as many as four sheaves in the top block and four in the bottom block will frequently experience so much friction that it is difficult to operate. This friction is caused mainly by the energy loss due to bending and unbending ropes.

A 'third system'

A 'third system' of tackle is arranged as shown below. The diagram at left is a schematic, the one on the right is more realistic, as will be seen shortly. The VR of the two-sheave set illustrated here can be deduced as follows:

Assuming for the moment that the link between ropes 'a' and 'b' and the load remains horizontal (*it doesn't: see below*): If the load moves up by one unit, then points 'a' and 'b' both move up by one unit. Due to point 'a' moving up by one unit, the centre of sheave (2) moves *down* by one unit. Stemming from this movement, point 'c' would move down by 2 units.

But, point 'c' also moves down by one further one unit to take up the slack from point 'b' moving up by one unit. Therefore point 'c' moves down a total of 3 units. VR = effort distance/load height raised = (3units)/(1 unit) = 3

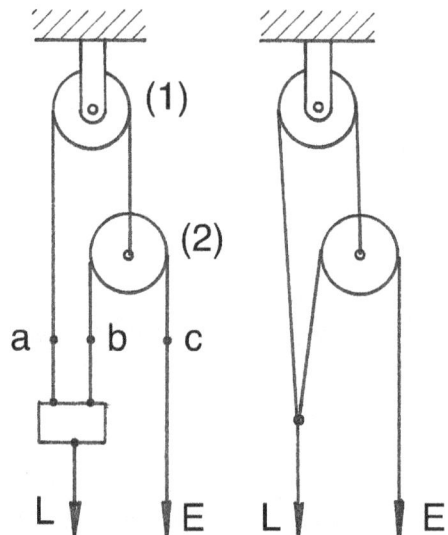

However, to get practical: a free-body diagram of sheave (2) will immediately reveal that the tension in rope 'a' must be be twice that in rope 'b', so the link on the left is *not* going to remain horizontal. That is why a more practical arrangement would be to join ropes 'a' and 'b' to the load rope directly, as in the right hand diagram above.

This 'third system' appears similar to the upside-down 'first system' illustrated previously, with one major difference: sheave (1) is now a standing sheave, whereas its counterpart in the 'first system' was a running sheave. This difference renders the VR of a 'third system' to have a value *one less* than the equivalent arrangement of an upside-down 'first system'. If you add successive sheaves in a progression to this set, instead of having the VR increase in a sequence of 4,8,16, 32 and so on, it will increase in the sequence 3,7,15,31 and so on.

As was the case for a 'first system', going beyond two sheaves introduces various practical difficulties, making this arrangement unwieldy to set up and use.

Exercises: Verify the stated values for the VR of each of the following SLMs

4.732; 100; 4.8; 7; 10

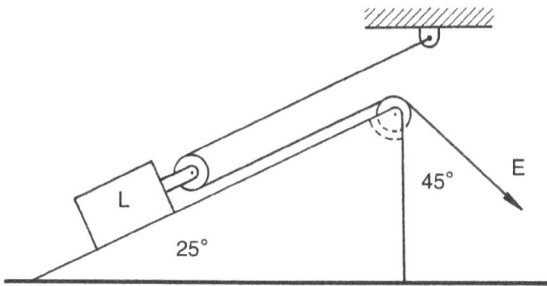

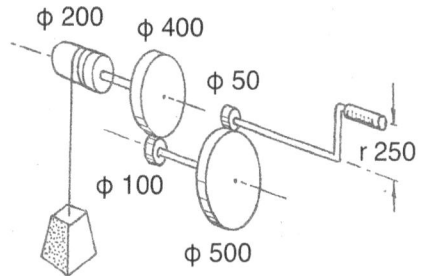

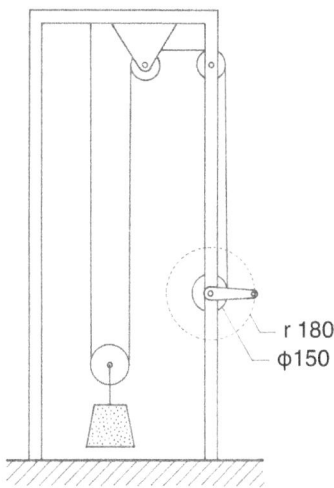

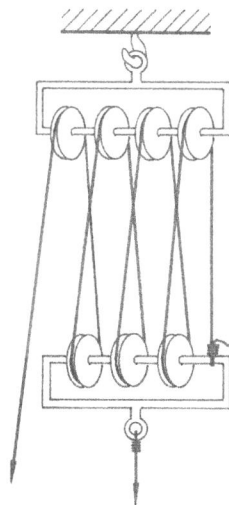

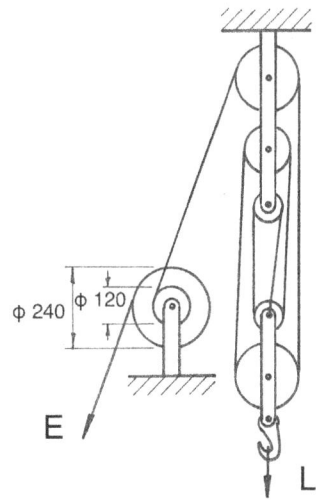

The 'law' of a machine

A graph of effort, E, vs. load, L for any lifting machine, will nearly always be a straight-line graph that looks as follows, whose equation is of the form

E = mL + C

where m = the gradient of the graph, and C is the value of the y-intercept, namely the effort required to operate the machine with zero load. Since this equation describes the effort-load relationship, it is known as the 'law' of the machine.

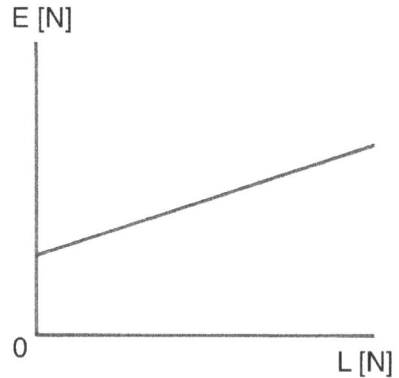

If we insert into this expression the numerical values of 'm' and 'C' that apply to a given machine, we have an equation that allows us to predict the value of effort required to raise any given value of load.

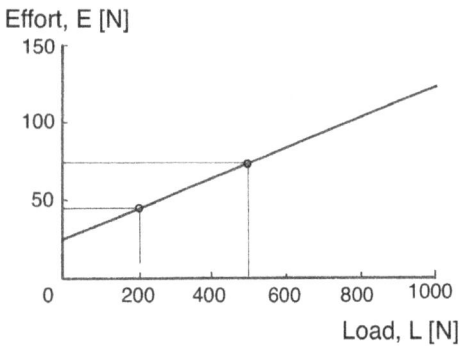

Example

For a given lifting machine, when the load is 200 N, the effort required is 45 N. For a load of 500 N, the effort needs to be 75 N. Draw a graph of effort vs load, to scale, deduce the 'law' of this machine, and use the 'law' to predict the effort needed to raise a load of 1 kN.

Using the data provided by the co-ordinates of the given points:

gradient: m = (75 N – 45 N)/(500 N – 200 N) = 0.1

Substituting this value into the general equation E = mL + C, and using the simultaneous values for one of the given co-ordinates:

75 = 0.1(500) + C   ∴ C = 25 N

The 'law' of this machine is thus **E = 0.1L + 25**   Hence, if L = 1000 N, E = 125 N

Exercise

In a certain lifting machine, the point of application of the effort moves five times as far as the vertical height that the load is raised. It is found that an effort of 14 N is required to raise a load of 20 N, and an effort of 26 N is required to raise a load of 80 N

a.   Sketch the graph of effort vs. load for this machine.

b. Deduce the 'law' of this machine, namely an expression relating effort E to load L.

c. Determine the mechanical advantage of the machine when a load of 200 N is raised.

d. Determine the velocity ratio of this machine.

[E = 0.2L + 10; 4.762; 5]

Why the mechanical advantage is always less than the velocity ratio

We saw earlier that in the ideal case, where no energy was lost during the operation of a lifting machine, the MA would be equal to the VR. However, we also know that it is inevitable that some energy will be lost. This means, that, realistically, in order to raise a load, the person applying the effort will have to do *more* work than is needed simply to raise the load.

Since work done = force × distance, in order for the operator to do a greater amount of work, he/she will either have to exert a greater effort, or exert the effort over a longer distance than before. However, the second of these two options is not usually possible, because the ratio of effort distance to load height raised is *fixed* by the VR, which is a constant characteristic of a given machine. So, the only practical way of putting in more energy is to exert a *larger* effort force.

Since the MA = (load)/(effort), if the effort is made larger, the value of the MA will be smaller. Hence, the MA is always less than the VR. The amount by which it is less gives an indication of the efficiency of the machine.

The efficiency of a lifting machine

The efficiency of a lifting machine is defined as the ratio of useful work done on the load to the work done by the effort.

Efficiency e = (useful work output)/(work input)

= (work done on load)/(work done by effort)

= (load × height load raised)/(effort × effort distance)

= (L × h) ÷ (E × d) = (L/E) ÷ (d/h) = MA ÷ VR

For example, if a SLM had a VR = 10 and it was found that for a given load, the value of the MA was 8, the efficiency of that machine while raising that particular load, is given by 8/10 = 0.8 or 80%.

The efficiency is usually expressed as a percentage. 100% efficient means there are no energy losses. 80% efficient means that only 80% of the input energy went into useful work to raise the load, and that the other 20% of the input energy was unrecoverable.

Why the efficiency of a simple lifting machine varies with load

Consider the energy accounting diagrams for three separate operations of a given lifting machine, firstly, with a small load, secondly, with a larger load, and thirdly, with an even larger load.

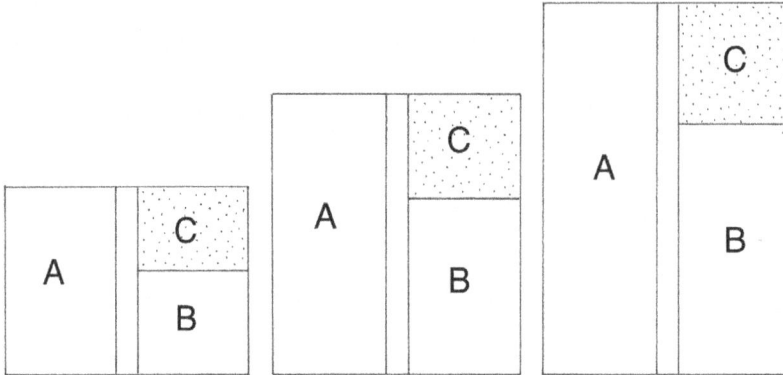

Since efficiency, e = (work done on load)/(work done by effort), and these two amounts of work are represented respectively by the heights of columns B and A shown on the diagram, we have: efficiency e = B/A, in each case.

The height of column C represents energy losses, which may be assumed not to increase *significantly* for greater values of load. Column C comprises friction losses as well as work done to raise machine parts. There is definitely *some* increase in the energy losses with load, as the work done against friction will increase, and the relatively minor amount of work done to stretch ropes will also. However, the work needed to raise machine parts will remain constant.

As the load increases, quantity C, which does increase, but not proportionately, becomes less significant when compared with B. Therefore, as the load increases, the ratio of B/A becomes greater. Hence, the efficiency increases.

The above diagram is drawn to scale. The efficiencies in this particular diagram, represented by the ratio B/A in each instance, are respectively 55%, 63% and 68%. The figures are typical for a SLM with respectively light, moderate and high loading.

The reasoning described above is reflected in the shape of an actual graph of efficiency vs. load, plotted from experimental data, which looks as follows:

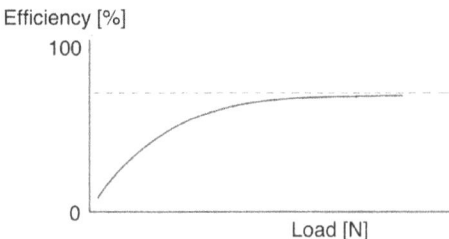

The graph of efficiency vs. load approaches an asymptote, whose value represents the maximum possible efficiency of that machine.

The maximum load that can be raised is limited by the strength of the machine, so the load cannot be increased indefinitely.

Hence, it is unlikely that the efficiency of any machine will ever reach its maximum possible value. Note that the value of maximum possible efficiency differs from machine to machine.

Determining the maximum efficiency of a SLM

Consider the general law of a machine, namely $E = m L + C$. From this equation,

$L = (E/m - C/m)$. Now, divide through by E:

$L / E = 1/m - C/Em$ and, since L/E = the MA, this becomes

$MA = 1/m - C/Em$. Now, as the load increases, so does the effort, E, so that at very high values of load, C/E becomes very small. In the limit, this value tends to zero, so the maximum value of the MA is effectively $(1/m)$.

Since efficiency, $e = MA /VR$, the maximum efficiency,

e_{max} = (maximum MA)/VR = (1/m)/VR = 1/(mVR)

Example

A particular simple lifting machine has a velocity ratio of 8. In order to operate this machine with zero load, an effort of 54 N is required. If the load is 1000 N, the effort needed is 254 N.

Determine the maximum efficiency of this machine, the maximum MA of this machine, and the efficiency of the machine when the load is 1000 N.

$E = mL + C$ $\therefore 54 = m(0) + C$ ....(1) and $254 = m(1000) + C$ ....(2)

From (1), C = 54 N. Substituting this value in (2) yields: m = 200/1000 = 0.2

So, the maximum efficiency of the machine:

e_{max} = 1/(mVR) = 1/(0.2 x 8) = 0.625 or 62.5%

The maximum MA = 1/m = 1/0.2 = 5

When the load is 1000 N, The MA = L/E = 1000/254 = 3.937

Efficiency at this load value, e = MA/VR = 3.937/8 = 0.492 or 49.2%

A safety consideration in lifting machines

Under what circumstances will a lifting machine run back, or 'overhaul', namely operate in reverse, when the effort is removed?

Certain lifting machines pose a danger to the operator and the surroundings: for example, when raising a load using a block and tackle, if you accidentally let go of the effort rope, the load will descend dangerously fast, seemingly without resistance.

The same would happen if one let go of the crank handle of a winch. That is why winches and hoists are often fitted with a self-activating brake, or a pawl and ratchet: to prevent such an accident.

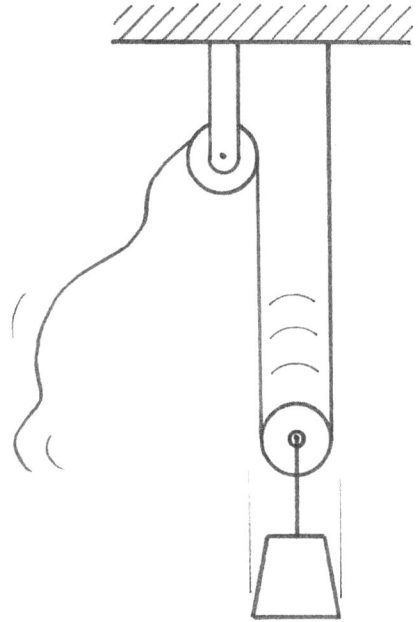

A winch is identical in construction to a hoist, except that it is usually used to pull horizontally. Below is an illustration of a winch with ratchet and pawl.

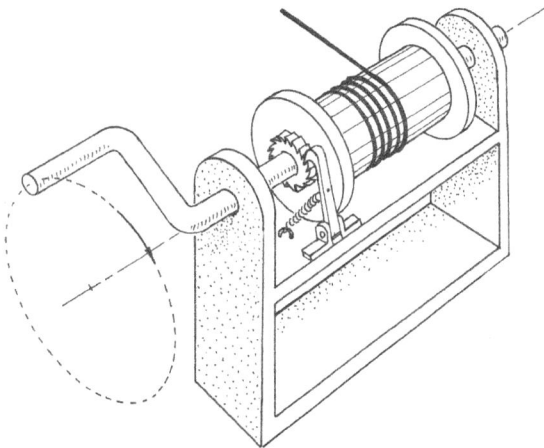

Accidental running back of the load does not occur, however, when raising a motor car with a screw jack, or when pushing a crate up a ramp. If one stops applying the effort force, these loads will simply stay in place. Why?

Whether or not a machine will 'overhaul' depends on the amount of friction that has to be overcome in order for the machine to reverse.

If the work that needs to be done against friction in order to operate the machine in reverse is greater than the potential energy already gained by the load, then the load can't descend. This phenomenon is illustrated by examining the energy accounting diagrams of two different machines, shown below:

Firstly, consider a machine whose efficiency is greater than 50%.

Work done by effort	A	C	Work done to overcome friction	
		B	Work done to raise load	

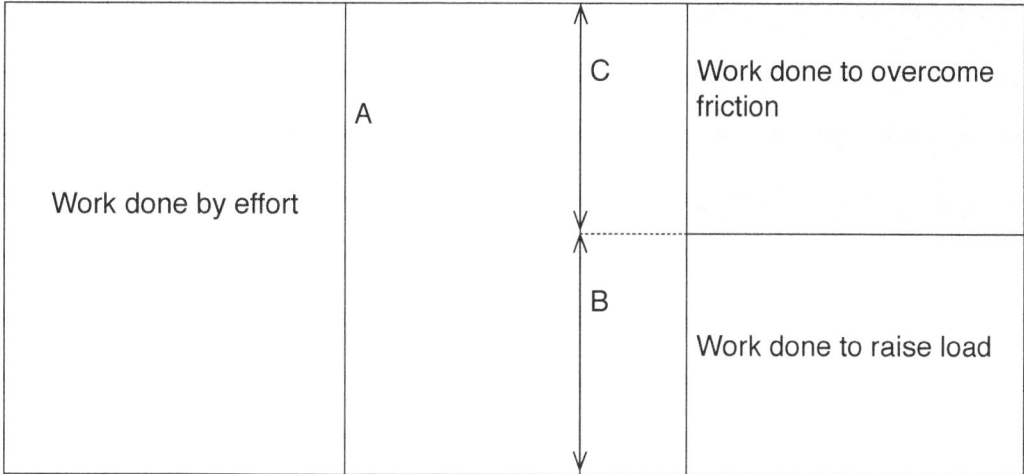

The efficiency of this machine: $e = B/A$

If $B > C$, then $B > 0.5\,A$, in other words, the efficiency is greater than 50% .

In this case, the potential energy possessed by the load (imparted to it by the work done on the load) is greater than the work needed to be done against friction to operate the machine. Therefore, if the effort is released, the machine will run backwards, allowing the load to descend.

Secondly, consider a machine whose efficiency is less than 50%.
Here, B will be smaller than C, and the energy accounting diagram looks as follows:

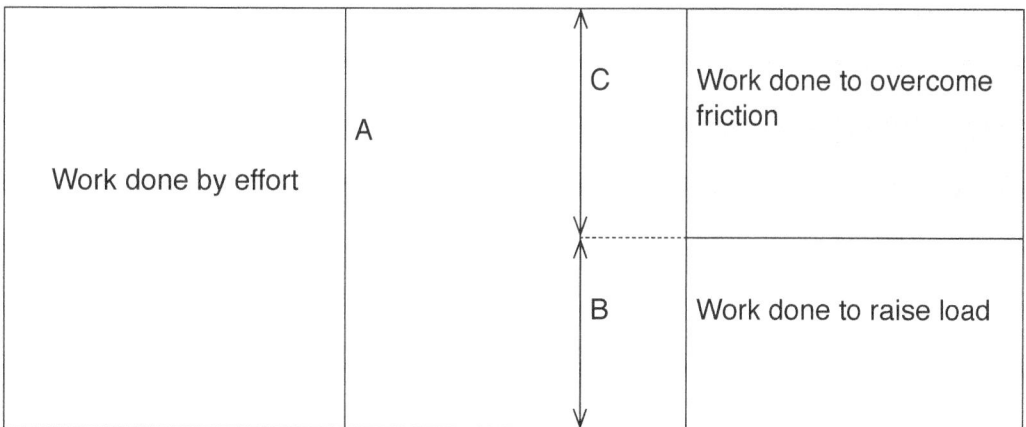

Work done by effort	A	C	Work done to overcome friction	
		B	Work done to raise load	

In the latter case, if the operator lets go of the handle, the amount of potential energy 'B' already imparted to the load is insufficient to perform the work that would be needed to overcome the friction, so the machine will *not* overhaul. The load will not descend, and the lifting machine is safe.

With most lifting machines, it is desirable to have a high efficiency, in order not to waste energy when operating the machine. However, in cases where it is important, for safety reasons, that a machine should *not* overhaul, the efficiency of the machine should be kept lower than 50%. This is accomplished by having an arrangement that deliberately provides a lot of friction.

A notable instance of this is found in lifting jacks that operate by means of a screw thread.

The efficiency of a screw jack is *designed to be* less than 50%. It has to be, for a very good reason. If you let go of the handle, you want the jack to remain in position. It should not run back to the starting point. If it did, the heavy load you are raising would descend rapidly, with dangerous consequences, and even before it landed, the crank handle would spin rapidly in the opposite direction to which it was turned, presenting a danger to the operator.

Consequently, a motor car jack that operates by a nut on a threaded bar is very inefficient, in terms of energy needed to raise the car. If it is 20% efficient, for example, you have to put in *five* times the energy that is actually needed to increase the gravitational potential energy of the car, to raise it to the desired height. This fact likely to be confirmed by the experience of any person who has had to crank such a jack. However, you can rest assured: a screw jack will not run down unless you crank it down.

Specific characteristics of several types of SLM.

The screw jack

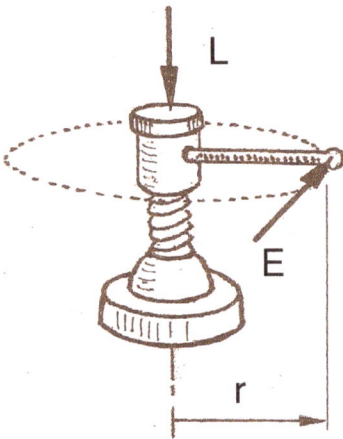

We have previously derived an expression for the VR for a screw jack: $VR = 2\pi r/p$ where p is the thread pitch.

Since the work done to raise the load for one full turn of the nut is $L \times p$ and the work done by the effort is $E \times 2\pi r$,

the efficiency, e = (work done on load) ÷ (work done by effort) $\therefore e = (Lp) / (E2\pi r)$

Consider some typical figures in these equations: firstly, using the equation for the VR: suppose the screw thread pitch is 5 mm and the rotating arm is 300 mm long. The VR of this jack would be $2\pi(300)/5 = 377$.

This means that to raise the load by 100 mm, one would have to move the end of the rotating arm 377 x 100 mm, which is 37.7 m, representing 20 full turns of the nut. If you have ever used a screw jack of any kind, including a scissors jack, you will appreciate that you have to do a lot of turning before the load gets to the height you want it.

Secondly, using the equation for the efficiency: suppose the effort you can manage is 100 N (a reasonable value for a person to exert), and the efficiency is 40%. The load that can be lifted can be determined from: e = (Lp) / (E2πr)

∴ 0.4 = L(5) ÷ 100 × 2π (300)   ∴ L = 15080 N, and the MA will be L ÷ E = 150.8

The wheel and differential axle

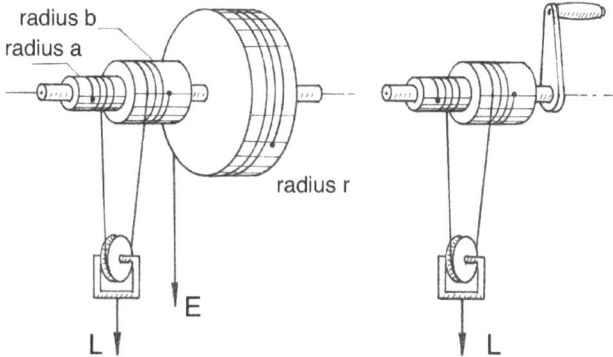

radius b
radius a
radius r
E
L
L

This machine is so simple that it is difficult to imagine one more elegant.

It consists of one solid moving part, which is a rotating assembly mounted in bearings, and one sheave, with a hook, that hangs in the loop of rope.

In operation, the rope is wound *off* the smaller axle and *onto* the larger one.

One turn of the wheel (see left diagram) or crank (equivalent arrangement on right) unwinds a length of rope from the smaller axle, equal to one circumference of that axle.

So the rope coming off the smaller axle moves down by a distance 2πa, and, at the same time, the rope winding onto the larger axle moves up by 2πb. The loop of rope is thus shortened by 2π(b − a).

Consequently, the load sheave moves up by half this distance, namely π (b − a). In the meanwhile, the point of application of the effort has moved a distance 2πr.

Since VR = (distance moved by effort) / (vertical distance moved by load), we have

VR = (2πr) ÷ [π (b − a)] = 2r ÷ (b − a)

By making the difference between 'a' and 'b' very small, one can make the VR very large.

For example, if the crank radius is 400 mm, radius 'a' = 100 mm and radius 'b' = 120 mm:

Then the VR = (2 x 400) / (120 - 100) = 800 / 20 = 40, thus enabling a person to raise a load 40 times the effort that can be exerted on the crank. The average person could easily apply a force of 100 N to the crank, and would thus be able to raise a load of 4000 N, roughly equivalent to the weight of five grown men.

Naturally, with this particular velocity ratio, one would have to move the crank handle a distance of 40 metres along its circumference of movement, in order to raise the load a height of one metre. This means that it will take quite a long time to raise the load.

Which fact brings us to consider one limitation of having a lifting machine with a high VR: the higher the VR, the greater the load you can lift, but the longer it will take to raise the load. If you have a very large VR, say, of 400, you would have to move the point of application of the effort a distance of 400 times the height that you wish to raise the load. This will test your patience, and is the reason why we generally avoid using a SLM that has a very high VR.

The chain hoist

Hand-operated chain hoists are frequently used in workshops to raise medium-sized loads. The particular design shown here has a compound sheave that rotates on a shaft. This shaft is usually supported by a trolley that runs on an overhead rail. The trolley and rail are not shown, as they are simply a means to move the lifting machine to where it is needed.

A chain hoist operates on the same principle as that of the wheel and differential axle.

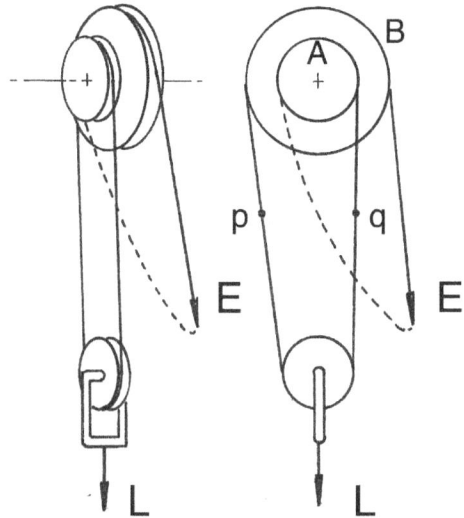

The chain that passes around the compound sheave is formed into an endless loop. The chain links lie in slots on the rims of the compound sheaves, so the chain cannot slip in the way that a rope might. The diameters of the compound sheave have to be chosen to house an integral number of chain links.

Suppose the number of chain-link slots in the rims of the two sheaves are as follows: sheave A has 'a' slots and sheave B has 'b' slots.

When the effort E turns the compound sheave through one revolution, point 'p' on the chain moves upwards by 'b' links, and point 'q' on the chain moves downwards by 'a' links.

The loop of chain in which the running sheave lies is therefore shortened by (b – a) links. Consequently, the load rises half that distance, namely (b – a)/2.

So, the VR = effort distance/height load raised = b/((b – a)/2) = 2b/(b – a)

The VR can be made very large, if the difference between 'b' and 'a' is made very small.

Example of a chain hoist calculation:

For a chain hoist of the type shown above, if the larger sheave has 40 chain-link slots and the smaller has 39 slots, determine the VR.

VR = 2b/(b – a) = 2(40) / (40 – 39) = 80

A chain hoist such as this one is a very effective SLM, but it must be remembered that, in order to raise the load by one metre, the operator will have to pull down continuously, hand over hand, 80 metres of chain.

How the tension in a rope varies throughout an arrangement of tackle.

Consider a rope passing over a fixed drum, carrying a load at one end, and being pulled in the direction of an effort force at the other end. The phenomenon is easily demonstrated by using a piece of rubber band like a rope to pull a light load over a fixed horizontal rod.

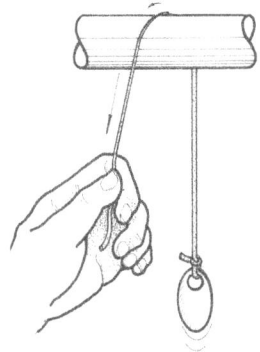

As the rope slides around the surface of the drum, the friction between the rope and the drum will cause the tension in the leading rope to be greater than that in the trailing rope.

There is a particular relationship between the tight side tension and the slack side tension in such circumstances. We won't go into that now. For the meantime, it is only necessary to realise that this tension differential develops.

The tension difference is present if there is *any* friction at all, even if the drum is able to rotate. This effect is also observed in arrangements of blocks and tackle. As long as there is *any* friction in a sheave, the leading rope will carry more tension than the trailing rope.

It follows, that for an arrangement of sheaves, the tension in the rope is greatest at the effort end of the rope, and diminishes progressively from one section of the rope to the next, until it is smallest in the part of the rope that is fixed.

If you set up a block and tackle using a woven nylon rope, such as a mountaineer's climbing rope, with a load of 30 kg or more, you can easily detect the progression of tensions, since the respective thicknesses of the sections of the rope, due to the successively increasing amounts of stretch in each section, are plainly visible.

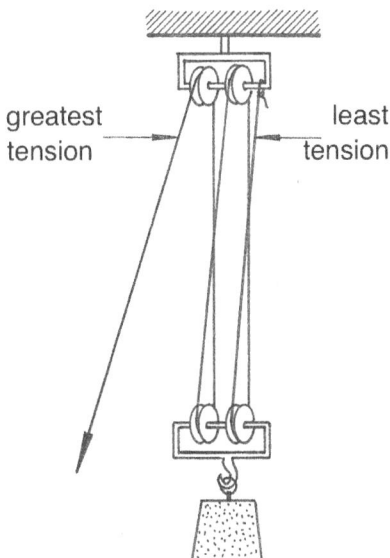

In the design of a block and tackle, this affects the choice of which load rating of rope to use. It is not sufficient to reason that since there are, for example, 4 ropes between the blocks, supporting a load of 4000 N, the rope should be rated for sustaining a tension of 1000 N.

The *average* tension in these four ropes will be 1000 N, but two of these ropes will have to sustain more than 1000 N, and the effort rope still more than that. A generous factor of safety is recommended.

The wedge as a lifting device

Mention was made earlier that the wedge is one of the earliest devices known to be used for lifting heavy loads.

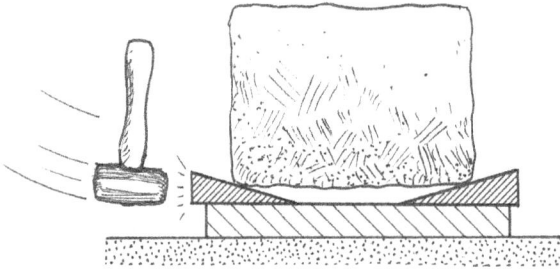

Can the use of a wedge be analysed in the same way as other lifting machines? Basically, no. There are too many unknown factors.

The value of the load might be known, but the effort is applied by the action of hammering the wedge, which is an impact situation, as opposed to a steady application of force.

This means that the value of the effort force cannot be determined, not even roughly.

Also, as a result of repeated impacts from hammering, the wedge itself, and the surfaces above and below it, are likely to become damaged, leading to an effective change of angle between the wedge and the surfaces it is separating, at the point of contact. If one was trying to do an analysis based on the wedge angle, this analysis would become inaccurate. The steeper the wedge, the more likely is such damage to occur.

The only criterion we can apply to the use of a wedge, is to determine the wedge angle at which slip would occur, assuming that no deformation of materials takes place. From this we can deduce which wedge profile would hold a load safely without slipping.

Recall, from the chapter on friction, the angle of repose, ϕ (phi), which is the angle to which a plane surface must be raised for a load placed on the plane to slip under its own weight. For a load resting on a wedge to slip down the angled surface of the wedge: the wedge angle must be greater than or equal to the

Profile A: $\mu = 0.1$ $\phi = 5.71°$

Profile B: $\mu = 0.2$ $\phi = 11.3°$

Profile C: $\mu = 0.3$ $\phi = 16.7°$

Profile D: $\mu = 0.4$ $\phi = 21.8°$

Profile E: $\mu = 0.5$ $\phi = 26.6°$

angle of repose, which is related directly to the coefficient of friction, μ, by the relation $\phi = \tan^{-1}\mu$.

The accompanying diagram shows some values for μ, together with the equivalent values of the angle of repose, and a diagram of each angle drawn to scale, to represent the profile of a wedge made to that angle.

The implications for the holding stability of a wedge, according to the coefficient of friction between wedge and load, are as follows:

For teflon on steel ($\mu = 0.1$): a wedge of profile A (or sharper) will be required to hold without slipping.

For nylon on nylon ($\mu = 0.2$) and hardwood on steel ($\mu = 0.2$), a wedge of profile B (or sharper) will hold without slip.

For wood on wood ($\mu = 0.4$), and *all* other combinations of material, a wedge of profile D (or sharper) will hold. In some instances, where μ values exceed 0.5, wedges of profile E and even steeper profiles might be able to hold without slip. For most applications, however, wedges of angle 20°, which is slightly sharper than profile D, will hold any load reliably.

Exercises: all aspects of simple lifting machines

Question 1

The block and tackle shown here is used to lift a load of 800 N. The effort force, E, required to do this is found to be 448 N. When the load is 1200 N, an effort of 608 N is required.

- How far must the effort rope be pulled downwards, in order to raise the load by 2 m? [4 m]
- What is the mechanical advantage of the machine when raising this load? [1.786]
- What is the efficiency of the machine at this value of load? 89.29%
- Determine the 'Law' of this machine. [E = 0.4L + 128]
- Use the 'Law' to predict the effort needed to raise a load of 1500 N. [728 N]

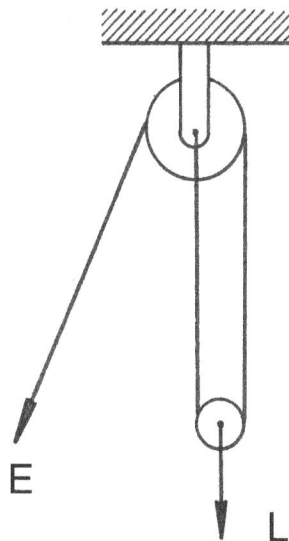

Question 2

The geared hoist illustrated here is used to raise a load L. Crank A is fixed to gearwheel B, which meshes with gearwheel C. The winch drum D is fixed to C.

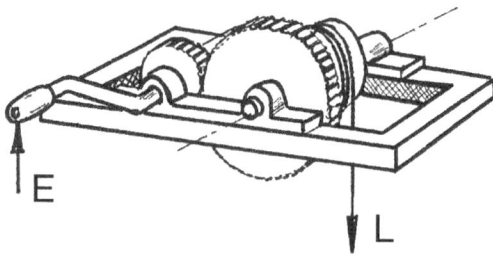

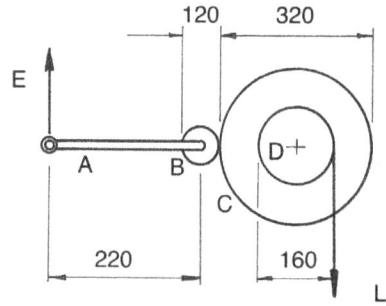

- Deduce the velocity ratio of this machine, explaining all the steps of your reasoning. [7.333]

- If no energy were lost to friction, what effort would be required to raise a load of 1060 N? [144.5 N]

- Would you expect the efficiency of this machine to be high, or low? Explain. [high]

Question 3

Consider the following lifting arrangement, combining sheaves and an inclined plane:

The box with the sheave attached weighs 10 N. The coefficient of friction between the box and the inclined plane is 0.2. Assume no friction in the sheaves. The cord is light and can be presumed not to stretch.

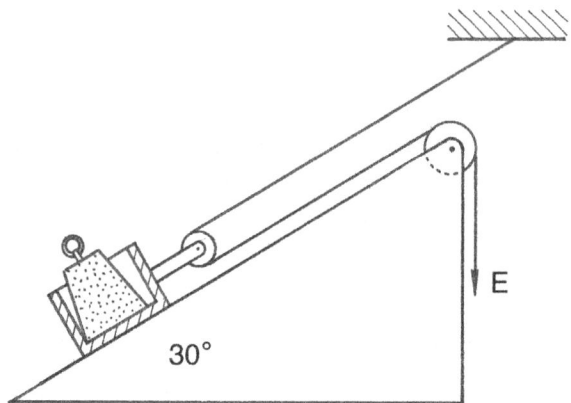

- Determine the velocity ratio of this lifting machine,.

- Establish the 'Law' of this machine, namely an expression relating effort E to load L.

- Determine the mechanical advantage and the efficiency of this machine for loads of 100 N and 1000 N respectively.

- Draw up an energy accounting diagram for this machine, for the operation of raising a load of 1000 N, through a vertical height of 1 m.

[4 ;   E = 0.3366L + 3.366 ;   2.70 and 67.5%;   2.94 and 73.5% ;   1359.9 J, composed of 1000 J + 349.9 J + 10 J]

Question 4

Consider this schematic diagram of a geared hoist.

The axis of the load drum and the crank axis are not in the same straight line. The number of teeth on each of the two smaller gearwheels, A and C, is 20.

Determine suitable numbers of gear teeth for the two large gearwheels, to enable an operator to raise a load of 800 kg.

Gearwheel B must have at least 1.5 times the number of teeth as gearwheel D. Assume the efficiency of the hoist is 96% at this load value.

The operator can exert a tangential force of 120 N on the crank. [144 and 96 teeth]

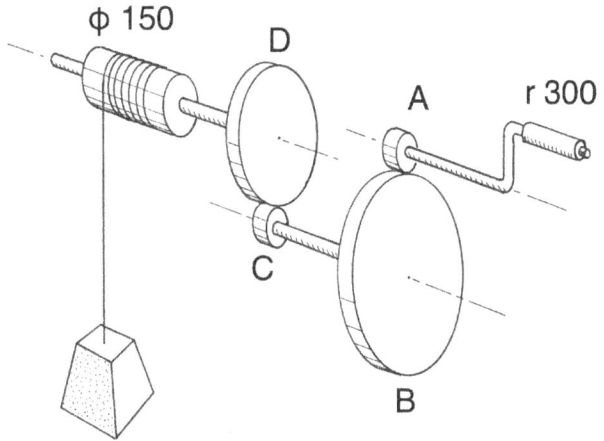

Tutorial questions for solving in small groups.

To be solved by collaboration: answers not provided here.

Tutorial Question 1:   Consider the four different arrangements of blocks and tackle shown schematically here:

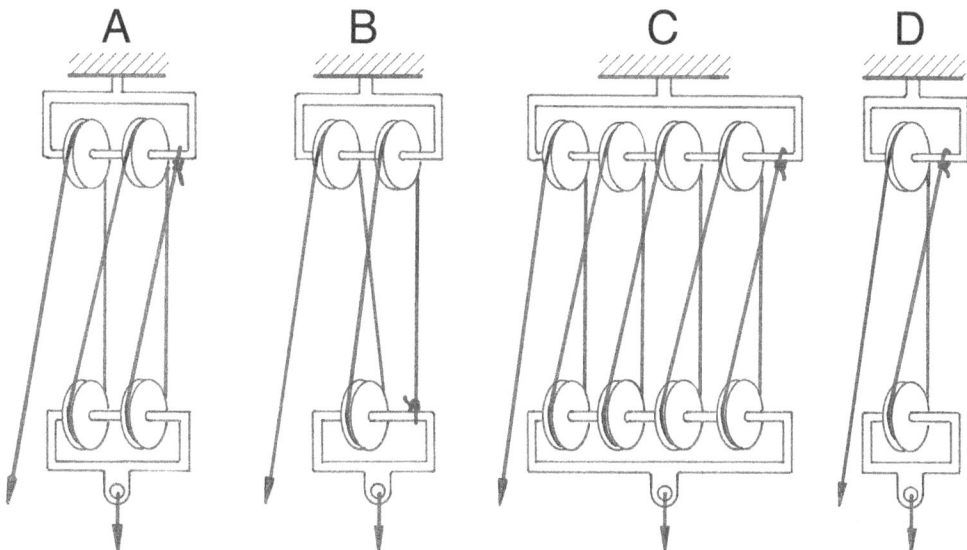

1. State the VR of each arrangement. A............B............C............D............

2. If all the individual sheaves exhibit the same amount of friction, which arrangement would be the most efficient? ....

3. For every metre that the respective loads are raised, how many metres would one have to pull the respective effort ropes? (assume no stretch in the rope) A....................B........................C........................D....................

4. If the loads were all equal, which arrangement would require the least effort to raise the load? ....

5. Suppose the friction in each of the sheaves causes a tension difference of 5 N between the leading and trailing ropes, and that each of the lower blocks (not including sheaves) weighs 80 N, and that each sheave weighs 10 N. For each arrangement, determine the effort, E, required for the given values of load, L, and complete the table below:

Load, L [N]	Effort, E [N] for the respective arrangements			
	A	B	C	D
200				
400				
600				
800				

6. Use data from the table above to plot a graph, to scale, of E vs. L for arrangement C.

7. Determine the 'law' of the machine for arrangement C.

8. Determine the efficiency, e, for each of the four given values of load.

9. Plot a graph of efficiency, e, vs. Load, L for this machine (arrangement C).

10. Determine the efficiency that one could expect if L = 1000 N

11. Looking at your graph: does it appear that this machine would have a maximum efficiency? If so, draw in the asymptote, judging by eye where it ought to be. What is the value of the maximum efficiency?

12. Compare the above value with the value you would obtain from the equation for maximum efficiency, namely $e_{max} = 1/(m \, (VR))$.

Tutorial Question 2

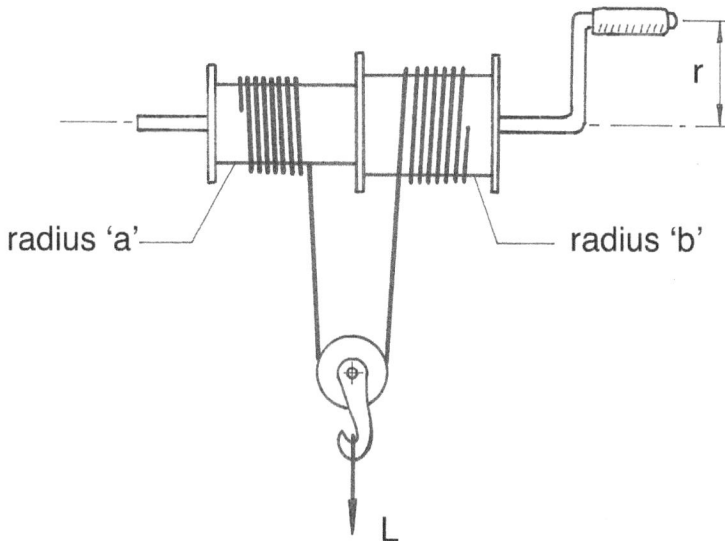

1. Suppose you need to construct a crank and differential axle, where radius 'a' = 160 mm and the crank radius r = 280 mm. If you want it to have a velocity ratio of 50, determine a suitable dimension for radius 'b'.

2. If you operate the machine at such a speed that it takes 2 seconds to turn the crank one full revolution, how long would it take to raise a load through a height of 0.5 m?

3. If you operate the machine at the above speed, when the load is 1000 N, what is the required power input from the operator? (Assume no energy losses due to friction, and that the hanging sheave with its hook weighs 25 N.)

4. If you wind the crank like crazy and move it one full turn every 0.5 seconds, when the load is 1000 N, what would be your power input to the machine?

5. In raising the load through 0.5 m, how many turns of rope are wound off the smaller axle onto the larger one? (Regard one turn as one circumference of the smaller axle.)

6. If the rope is 5 mm thick, how long should the smaller axle be, so that all the rope comes off the axle directly (without it being necessary to have some of the rope wound over previous turns of the rope)?

7. For a load of 1000 N, what is the MA?

8. For a load of 1000 N, what is the efficiency, expressed as a percentage?

9. If the maximum force you can exert on the crank is 80 N, what is the maximum load could you lift, using this machine?

Advanced Exercises: Designing simple lifting machines

Advanced Exercise A

Suppose you need to design a block and tackle arranged as a 'second system' for one operator to raise a load of 1500 N a height of 2.5 m. For a first estimate, assume the operator is comfortable exerting a downward force of 200 N. The pulley blocks available to you all have sheaves of diameter 140 mm, and weigh as follows:

1-sheave block.... 30 N   2-sheave block.....40 N   3-sheave block....50 N

4-sheave block.... 60 N   6-sheave block.... 80 N

Suppose that each sheave contributes a frictional resistance equivalent to adding 5 N of tension to the rope. Assume there is no stretch in the rope.

    a. What VR is required?

    b. How many sheaves would you need in the top block, and in the lower block?

    c. What is the minimum length of rope you would need?

    d. What is the average tension you would expect to have in any part of the rope?

    e. What tension could you expect in the effort rope? Could the operator manage this?

    f. If the operator can work at a rate of 200 W, how long would it take to raise the required load the full distance?

    g. In one operation of this block-and-tackle, how much work will have to be done in raising machine parts?

    h. What is the efficiency of this machine when raising the given load?

[a. VR = 7.5 hence 8 ropes, hence actual VR = 8   b. top block 4 sheaves, lower block 4 sheaves.   c. minimum 25 m, preferably at least 30 m   d. 200 N   e. 240 N   f. 24 seconds   g. 150 J h. 78.1%]

For the following questions, exact answers cannot be provided, because the given data is insufficient to solve these exercises completely. Some additional reasonable estimates have to be made. Common sense and experience are needed to make these necessary estimates.

Advanced Exercise B

You need to construct a hand-operated geared winch to raise a load of 200 kg through a height of 2m. Assume that this machine would have an efficiency of 85%.

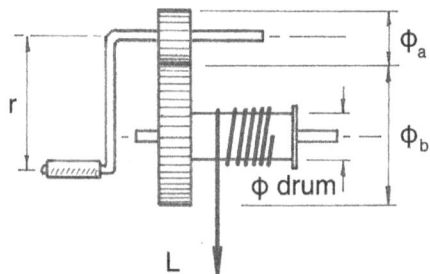

First, provide reasonable estimates for the following values:

The tangential force you can exert on a crank handle,

The rate at which you can expend power, and

A suitable radius for the hand crank.

Then, using the above values, determine suitable values for:

a. The overall velocity ratio
b. The pitch circle diameters of the pinion and gear respectively, given that each gear tooth will occupy 15 mm along the pitch circle circumference
c. The drum diameter
d. The number of seconds it would take you to raise this load the required distance.

Advanced Exercise C

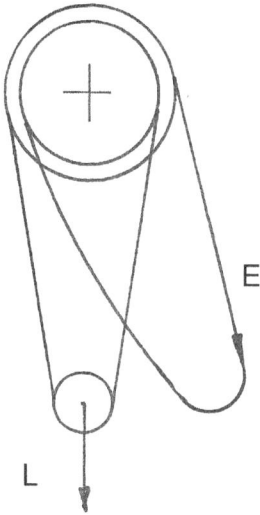

In designing a chain hoist of the type described earlier in this chapter, supposing there are two types of chain available. Type 'A' has links 20 mm long, weighs 1.5 kg per running metre, and has a load rating of 400 kg. Type 'B' has links 30 mm long, weighs 2.8 kg per running metre, and has a load rating of 800 kg. Assume the efficiency of this machine will be 80%.

This hoist has to be used to raise loads of up to 720 kg, through heights ranging up to 4 metres from the floor.

a. Would both types of chain be suitable? Explain.

b. Which type would you use? Why?

c. Assuming the operator can exert a maximum comfortable downward force of 120 N, decide on suitable diameters for the compound sheave, and for the running sheave.

d. When raising a load of 720 kg, with what force will the operator need to pull down on the chain?

e. If the load needs to be raised through a total height of 500 mm, what length of chain will have to pass through the operator's hands?

f. If the compound sheave is mounted on a travelling rail, so that its centre is 4.0 m above the floor, what would be a suitable length of chain to use from which to make the endless loop?

135

A design-and-build project on the topic of SLMs:

As with most phenomena in engineering, you won't fully comprehend simple lifting machines until you have played with them, done experiments with them, or, better still, built one and tested it.

This author frequently made use of design-and-build projects so that his students would get to grips with this topic.

One such project, for example, required students, in groups, to build a geared hoist, almost entirely of wood. The hoists were tested to destruction on a testing jig. Each machine was progressively loaded until it failed. Each project earned a performance index (P.I.), calculated as follows:

P.I. = (ultimate load at failure) ÷ (weight of the geared hoist).

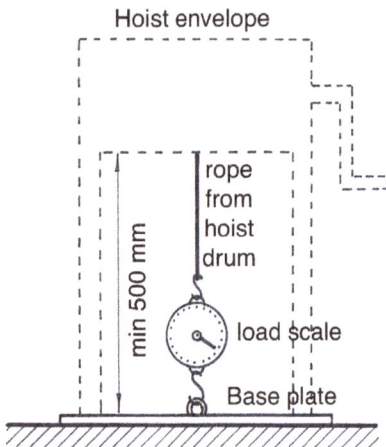

Groups competed with all other student groups to obtain the highest performance index for the project, as their marks would be directly related to this index.

There were specifications for the materials to be used, the manner in which the machines would be tested, and how marks would be allocated. All gears had to be made from wooden discs with pegs, in the manner of medieval gearing.

A diagram from the specification sheet is shown here.

Student resourcefulness and ingenuity were responsible for a wide variety of designs.

Examine the photograph of one student-built machine and see if you can identify:

- features of the design which are commendable,

- features which you think can be improved upon, and

- ways in which you could reduce the mass of the machine without compromising strength.

Use the following true/false test to check your understanding of simple lifting machines.

Identify whether each of the following statements is true or false:	T	F
1.	For a given SLM, the MA is usually greater than the VR.		
2.	The efficiency of a SLM is a measure of what fraction of the work done by the effort has gone into work done to raise the load.		
3.	For a block and tackle, the more sheaves it has, the higher the efficiency.		
4.	The smaller the load, L, the greater is the efficiency, e, of any SLM.		
5.	For the SLM shown here, if the load is 100 N, the effort would be 200 N. 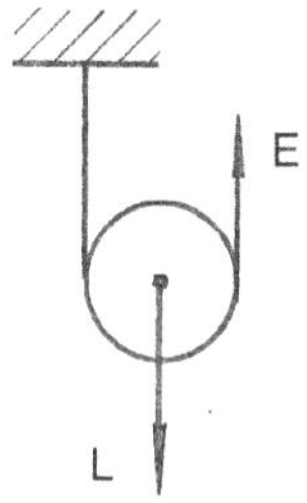		
6.	If the effort were removed during one operation of a machine that is 60% efficient, the load would descend.		
7.	The energy that goes into raising machine parts is available to help reverse the machine when the effort is removed.		
8.	If no energy were lost in the operation of a SLM, the MA would equal the VR.		
9.	The VR of this SLM is equal to 3. 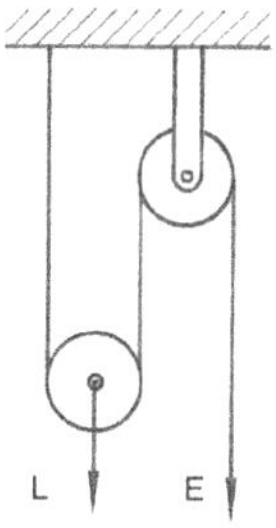		
10.	For this geared winch to have a VR of 10, the drum diameter 'd' should be 120 mm. 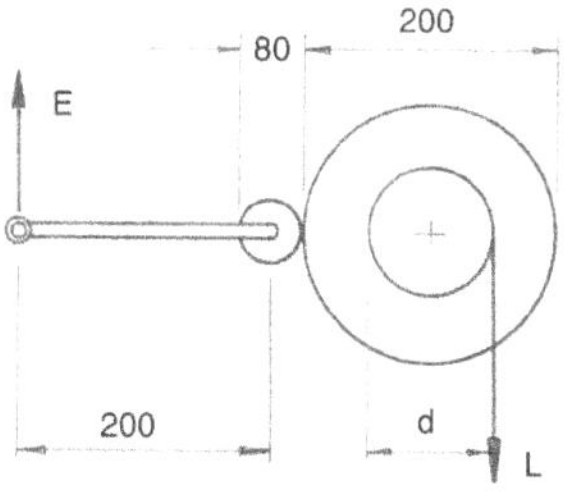		
11.	A graph of effort vs. load for a SLM cannot intersect the load axis.		

Chapter 15

Inertia in linear accelerating systems

The nature of inertia
Inertia is governed by Newton's 2nd Law
The acceleration of a single mass acted on by more than one force
The inertia of linked masses undergoing acceleration
The effect of the masses of the ropes that link the large masses

We have previously dealt with simple acceleration problems, in situations in which a single force acts on a single object. In this chapter we analyse situations in which a series of linked objects undergoes acceleration.

The simple analysis demonstrated here is important, because there are many instances in machinery and in transport where linked objects have to be accelerated or brought to rest.

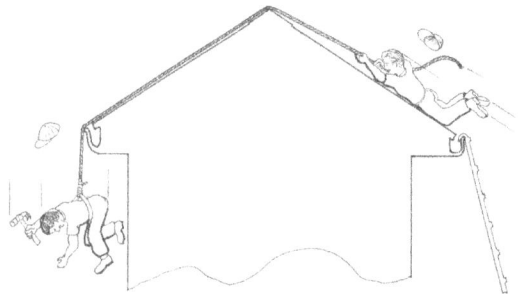

The analysis demonstrated in this chapter, while completely valid, is only part of what is needed to analyse the movement of linked objects.

The reason for the present analysis being only part of the picture is that most machinery has a rotating component as well as masses that move in straight paths. The effect of the rotational inertia of the rotating components of machines is significant.

If you have ever tried to spin a wheel by hand, you will know that the heavier the wheel, and the further away from the axis of its rotation that its mass lies, the harder it is to accelerate the wheel. This relationship is a manifestation of rotational inertia, which will be explained fully in a subsequent chapter, in volume 3 of this series.

For the present, we will not take into account rotational inertia, although it should be clear that where the rotating parts and the parts undergoing straight line motion are connected, the inertias of each of the linked components of both types affect the motion of the entire system.

In the present chapter, we will focus on the linear acceleration of such parts of a system of linked masses that are confined to movement in straight lines.

The nature of inertia

The more mass an object possesses, the more it resists changes to its state of motion. The property which resists such changes is called inertia. One can think of inertia as the tendency of an object to continue doing whatever it is doing. If it is standing still, it will take force to get it to move. If it is moving, it will take force to get it to come to rest. In both instances, the more mass it possesses, the greater is the force that is needed.

For example, when you add passengers to a car, it will accelerate slower than when there is only a driver in it. Also, once the car is moving, the more occupants it holds, the greater is the braking force needed to decelerate it at a given rate.

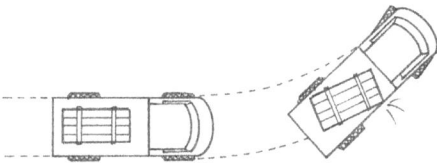

There are many examples of inertia to be found in daily life. For example, if you are driving along in a truck with an unsecured load in the tray, and take a corner, the load is inclined to continue in the direction it was travelling.

Stationary objects also possess inertia. If you propel a billiard ball to strike another ball in a direct line between their centres, the ball that is struck will behave in different ways, according to its mass. If it is a normal billiard ball, it will accelerate rapidly during the time of contact, and will move off with the same speed as that possessed by the approaching ball before the collision (*see chapter 16 on linear momentum in this volume*).

However, if it is a much heavier ball, say, made of steel, it will move off with a much smaller velocity. The more mass in the ball, the smaller will be its resulting acceleration.

Inertia is governed by Newton's second Law

One of the most basic knowledge patterns in mechanics is Newton's second Law, which tells us that when a force attempts to accelerate a mass, the acceleration that results is proportional to the force and inversely proportional to the mass.

The equation that relates these variables, is: **F = ma**

or, arranged differently: **a = F ÷ m**

A simple illustration of this relationship: suppose a motor-boat propulsion system is capable of exerting an effective driving thrust of 600 N. Ignore for the moment the resistance to movement offered by the water and the surrounding air. If the mass of the boat, including the driver, is 1200 kg, the acceleration from rest that is possible will be given by:

a = F ÷ m   ∴ a = 600 ÷ 1200 = 0.5 m/s^2

Now suppose this boat carries a passenger of mass 80 kg. Naturally it will accelerate at a smaller rate, and working again with $a = F \div m$, we see that the acceleration will be

$a = 600 \div 1280 = 0.4688$ m/s^2

When there is only one force and one overall mass involved, It is a straightforward matter to apply this equation.

The acceleration of a mass acted on by more than one force

Consider an object of mass 100 kg floating in space. Suppose two opposing forces act on it simultaneously: One force of 80 N to the right and another of 30 N to the left.

80 N ——————→ | **100 kg** | ←—— 30 N

The resultant of these two forces is 50 N to the right. We know that the resultant of a set of forces, being their equivalent, may replace the set. So, if we use the resultant instead of the individual forces acting on the object, we have in effect, the same situation as if there had been only one force acting on this mass.

To determine the acceleration, as before, $a = F \div m$ ∴ $a = 50 \div 100 = 0.5$ m/s^2

If one uses the term ΣF to represent the resultant of all the forces in the direction of the resulting acceleration, the equation for the acceleration may be written as:

$a = (\Sigma F) \div m$, or, when arranged in the original format in which this equation appears most commonly in the statement of Newton's 2nd Law: $\Sigma F = ma$.

Recall that equilibrium situations are governed by a similar-looking equation, namely $\Sigma F = 0$. That equation applied when the resultant of all the forces acting on an object was *zero*. It turns out that the equilibrium equation was a special case of the equation $\Sigma F = ma$, for a situation in which the value of 'a' is zero.

When applying the equation $\Sigma F = ma$, since we have to take account of the directions of all the forces that contribute to the magnitude of this resultant, it helps to recite in our minds the verbal interpretation of this equation, namely:

'The sum of the forces *in the direction of the resulting acceleration* equals ma'.

All the forces that are *in the same direction* as the resulting acceleration should be labelled as positive, while all those that are directly opposed to it should be labelled negative.

This simple acceleration equation, properly applied, will enable us to analyse systems of interconnected masses.

The inertia of connected masses

We need to be able to determine the linear acceleration of a system in which several masses are connected by ropes and are suddenly allowed to move, from rest.

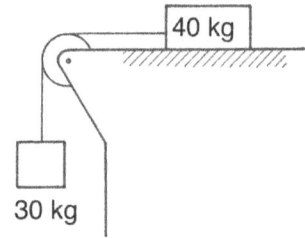

Example

A 40 kg stone block on a flat concrete platform is connected by a rope to another stone block of mass 30 kg that is suspended over the edge of the platform, but prevented from descending.

The rope passes over a sheave which may be regarded as light and frictionless. The coefficient of kinetic friction between the large block and the platform is 0.6.

If the 30 kg block is released, what will be the resulting acceleration of the system?

To apply the acceleration equation to any given mass-piece, we are interested only in the forces that act in the line of action of the possible acceleration of that mass-piece.

If we need to determine some of these forces from other considerations, we do that *before* applying the acceleration equation.

For example, we will need to know the friction force on the 40 kg stone, because the force in the horizontal part of the rope will depend on the value of this friction force. (Although it will not be *equal* to the friction force.)

The normal reaction between the platform and the block determines the magnitude of the kinetic friction force, which is given by:

$$F_k = \mu_k N = 0.6(40 \times 9.81) = 235.44 \text{ N}$$

The weight of the 30 kg block suspended from the rope is $30 \times 9.81 = 294.3$ N. Since this force > the friction force between the 40 kg block and the platform, it is likely that the 30 kg stone will descend, and the larger stone will slide to the left. However, we need to determine two unknowns: the force in the rope and the value of the acceleration.

The acceleration equation should be applied separately to *each* mass-piece. For this, we need to draw a separate FBD of each of the mass-pieces.

Since the two mass-pieces are linked, they will both experience *the same value of acceleration.* Let this value be 'a', and assume a direction for 'a'. The directions must be consistent, so that, if the upper block accelerates to the left, the lower one will accelerate downwards.

Draw in the acceleration vector on each separate FBD. To distinguish accelerations from forces, we will use a vector arrow with a double head for accelerations. Let the tension in the rope be 'T'.

The FBD of the stone block on the concrete platform:

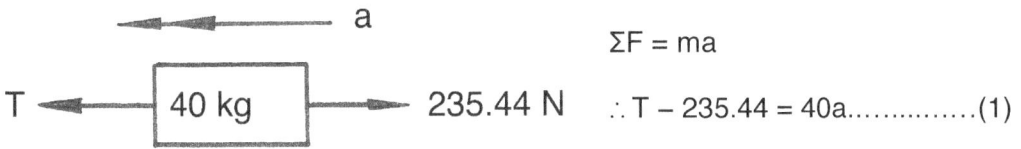

$$\Sigma F = ma$$

$$\therefore T - 235.44 = 40a \ldots\ldots\ldots\ldots(1)$$

(diagram labels: a, T ← 40 kg → 235.44 N)

The FBD of the suspended block of stone:

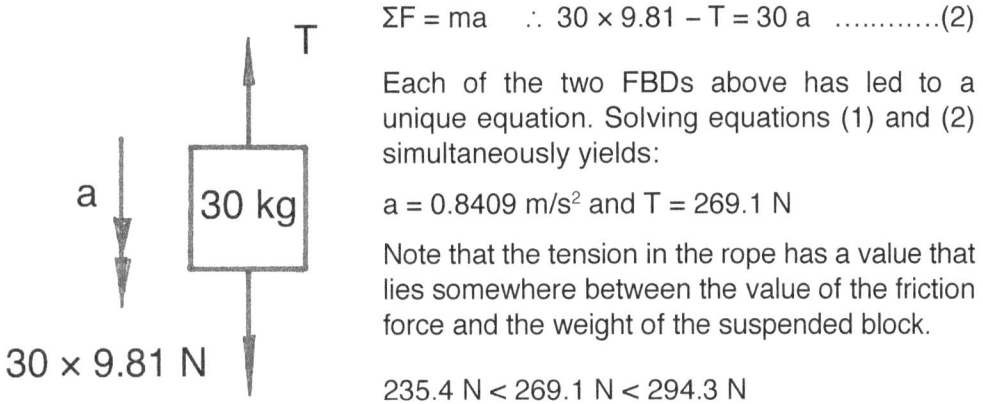

(diagram labels: T, a, 30 kg, 30 × 9.81 N)

$$\Sigma F = ma \qquad \therefore 30 \times 9.81 - T = 30a \ldots\ldots\ldots(2)$$

Each of the two FBDs above has led to a unique equation. Solving equations (1) and (2) simultaneously yields:

$a = 0.8409$ m/s^2 and $T = 269.1$ N

Note that the tension in the rope has a value that lies somewhere between the value of the friction force and the weight of the suspended block.

$$235.4 \text{ N} < 269.1 \text{ N} < 294.3 \text{ N}$$

This makes sense, for the following reasons:

1. The tension in the rope has to be *greater* than the friction force, or else the upper block would not move to the left.

2. The tension in the rope has to be *less* than the weight of the suspended block, otherwise the suspended block would not move downwards.

Such reasoning provides a useful check on any answers obtained for problems of this type.

Note also that the value of the acceleration of this system of connected blocks is very much less than the standard acceleration due to gravity. This is so on account of the resistances to the downward movement of the system. These resistances include the friction on the upper block, and the *inertia* of both blocks.

Exercises on the acceleration of linked masses

Question 1 *(see the following diagram)*

Two mass-pieces are connected by an inextensible cord passing over a frictionless sheave of negligible mass. The mass-pieces are initially restrained from moving, then suddenly released. Determine the tension in the cord and the acceleration of the system.

Reasoning from the preceding example, we can predict that the tension in the cord must have a value somewhere between the weight of the 20 kg block and the weight of the 30 kg block. This prediction can be used as a check on the reliability of our answer.

[T = 235.4 N and a = 1.962 m/s^2]

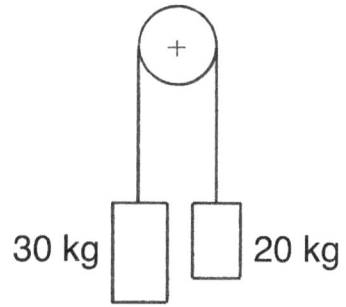

30 kg   20 kg

The low value of the acceleration in the above exercise is due to the inertia of the amount of mass that has to be accelerated. Here, both mass-pieces (totalling 50 kg) need to be made to accelerate, irrespective of the direction in which the pieces have to move.

Question 2

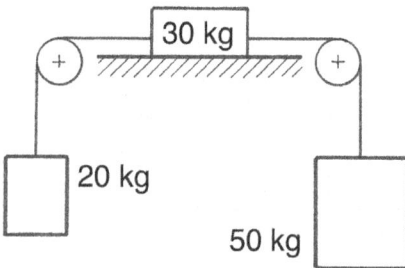

30 kg

20 kg

50 kg

Three mass-pieces are connected by two inextensible ropes passing over frictionless sheaves of negligible mass. The 30 kg block rests on a horizontal surface, on which the coefficient of kinetic friction between the block and the surface is 0.3.

If the 30 kg block is suddenly freed from an external restraint, determine the acceleration of the system and the tensions in the two ropes.

[a = 2.060 m/s^2;   T$_1$ = 237.4 N;   T$_2$ = 387.5 N]

Alternate method, useful as a check

For relatively simple layouts like those in questions 1 and 2 above, another way of determining the acceleration of the system is to evaluate the total force available to cause acceleration and the total mass that has to be accelerated, and apply the equation $\Sigma F = ma$ to these two quantities.

For example, in question 1, the total force available is the weight difference between the two sides, namely (30 − 20) × 9.81 N, and the total mass that needs to be accelerated is 50 kg.

Applying the equation $\Sigma F = ma$:   98.1 = 50 a   ∴ a = 1.962 m/s^2 which agrees with the value obtained by the method that considers each mass separately. The reader may wish to apply this method to question 2 to confirm its reliability.

However, while this method is useful as a check on the answer obtained by the previous method, this method does not provide values for the tensions in the ropes/chains that link the masses. If you want to determine those values, it is necessary to go back to the FBD of each individual item in the system and determine the forces acting on it, using the value of the acceleration that has been found by the short method.

Question 3

Two mass-pieces are connected by an inextensible cord passing over a frictionless sheave of negligible mass.

The mass-pieces rest on planes which are respectively at 30° and 60° to the horizontal. The 30 kg block is initially prevented from moving by a brake.

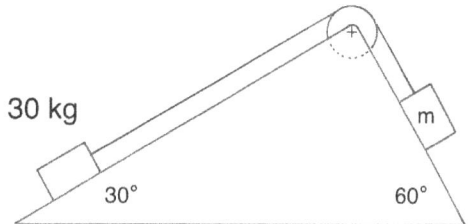

Determine the value of mass 'm' such that, when the brake is released, the system will accelerate at 0.25 m/s² to the right.   [m = 28.30 kg]

Question 4

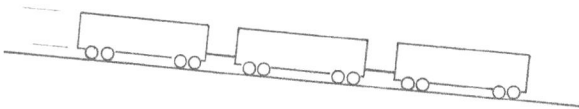

Three linked railway carriages are freewheeling down an incline angled at 5° to the horizontal. Each carriage has a mass of 30 tonnes and experiences a rolling resistance of 2 kN. When the speed of the train is 90 km/h, the emergency brake on the rear carriage is engaged, providing a braking force of 150 kN.

Ignoring the effects of air resistance, determine:

a. The value of the deceleration of the train.   [0.8783 m/s²]

b. The force in the link between the rear carriage and the middle carriage.   [100 kN]

c. The force in the link between the middle carriage and the front carriage.   [50 kN]

d. How far the train will move before coming to rest. [356 m, to the nearest metre]

Question 5

Two mass-pieces are connected by an inextensible cord passing over a frictionless sheave of negligible mass. The 80 kg mass-piece rests on a plane at 20° to the horizontal.

The 10 kg mass-piece hangs against a vertical wall, just touching the wall. The coefficient of kinetic friction on both interfaces is 0.2.

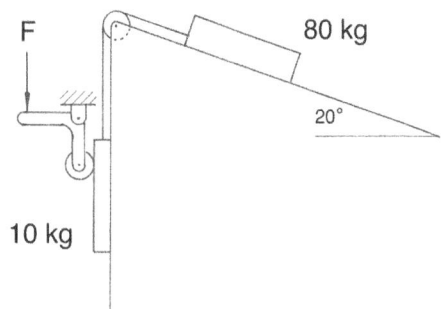

A roller can be pushed against the 10 kg mass-piece by applying a force F to a lever. The two arms of the lever are of equal length. Assume the mass of the lever to be negligible.

Determine the acceleration of the system:

a. If the roller is kept clear of the 10 kg mass-piece [a = 0.2536 m/s²] , and
b. If a force of F = 60 N is applied to the lever.   [a = 0.1236 m/s²]

Before doing any calculations, can you predict in which of the two situations, (a) or (b), the tension in the cord is likely to be greater? Is your prediction borne out by the calculations?

Question 6

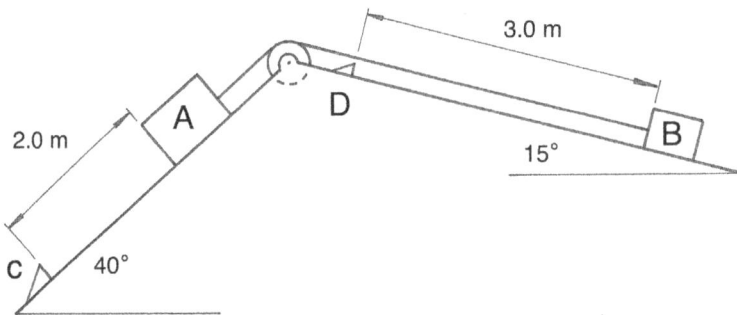

Two blocks, A (60 kg) and B (20 kg) are connected by a light inextensible rope passing over a frictionless sheave of negligible mass. The coefficient of kinetic friction is 0.3 for both blocks on their respective planes.

Initially, block B is held in place. When it is suddenly released, the system accelerates to the left until block A hits the stop at point C. Determine:

a. The initial acceleration of the system.   [1.693 m/s²]
b. The tension in the rope during this period of acceleration.   [141.5 N]
c. The velocity of block B at the instant that block A hits the stop.   [2.602 m/s]
d. The distance from stop D that block B comes to rest momentarily.   [369 mm]

The effect of the mass of the rope that links accelerating mass-pieces

In all the preceding examples it has been assumed that the tension in any one section of rope was the same throughout. However, this assumption is only valid if the rope has negligible mass, or if the rope itself is in equilibrium, namely, *not accelerating.*

A length of rope that possesses significant mass can only be made to accelerate if the tension at the leading end of the rope is greater than that at the trailing end. In this respect, the rope behaves like any other solid mass.

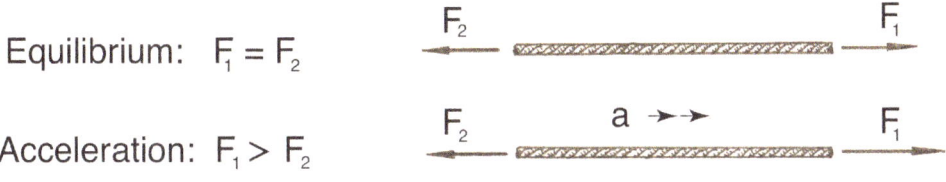

Equilibrium: $F_1 = F_2$

Acceleration: $F_1 > F_2$

If the mass of the rope is very much less than the mass of the objects that it links, the mass of the rope may be ignored, and this difference in tension between the ends of any section of rope may be ignored. However, if the mass of the rope (or chain) is significant, that difference in tension will also be significant.

In the types of example that we have encountered above, it is not easy to take account of the mass of the ropes, because in all these examples the rope passes over a sheave, so that an increasing length of the rope is moving in one direction away from the sheave, while a diminishing length of the rope is moving in another direction, towards the sheave.

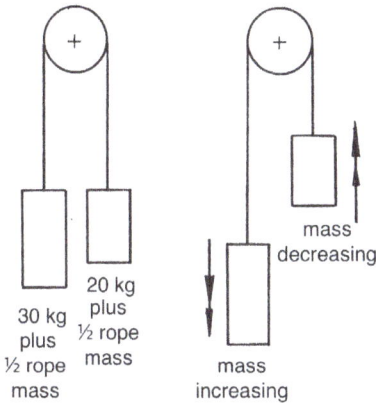

For example, in question 1 above, once movement begins, the part of the rope between the sheave and the 30 kg mass-piece will be lengthening, while the rest of the rope will be shortening.

This means that the changing distribution of the rope mass during the motion will cause the value of the acceleration of the system to *increase* progressively.

Such a situation is more complicated to analyse, and will not be dealt with here.

Using the acceleration equation as described in this chapter, we can, however, take into account the mass of the rope (or chain) in situations in which the whole of the rope or chain moves in such a way that all of its mass continues moving in the same direction. In such cases, the movement of the rope or chain does not introduce a variation in the acceleration of the system.

We can also rely on the acceleration equation to take into account the effect of the inertia of a chain in cases where there is an endless loop of chain. In such a situation, however much of the chain is

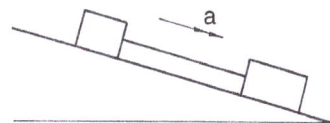

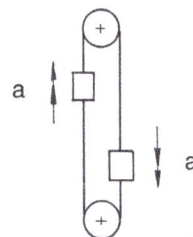

moving upwards, the same amount of the same type of chain is moving downwards at the same time. So we don't have a progressively changing value of acceleration.

Example

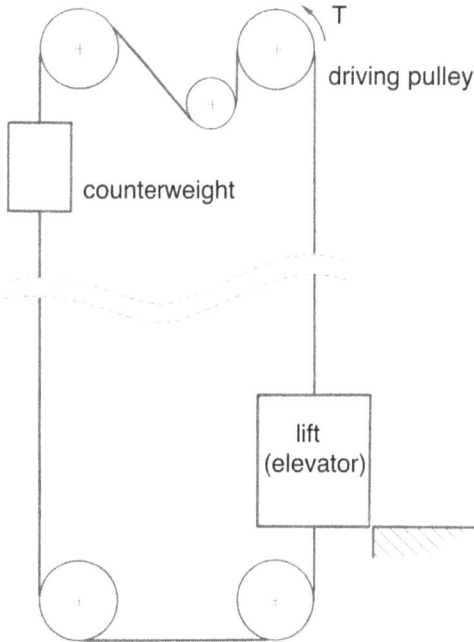

driving pulley

counterweight

lift
(elevator)

An elevator system consists of an endless loop made of 200 m of wire rope whose mass per running metre is 15 kg.

Attached to this rope are a lift of mass 600 kg and a counterweight of the same mass.

Determine the torque T needed to drive the system, firstly when the lift is empty, and secondly, with 3 passengers in the lift, of average mass 80 kg, such that the upward acceleration of the lift is 0.1 m/s².

The diameter of the driving pulley is 800 mm.

Ignore the rotational inertia of all the pulleys and assume there is no slip.

Solution: The total mass that needs to be accelerated is 4440 kg.

The force in the rope due to the torque T being applied is $F = T/r = T/0.4 = 2.5\,T$

Due to the presence of the counterweight, all other forces are balanced.

Applying the equation $\Sigma F = ma$ to the entire system: $2.5\,T = 4440(0.1) \therefore T = 177.6$ Nm

When the lift contains the three passengers: the force available due to the action of the torque on the driving pulley is counteracted by the weight of the three passengers, namely 2354.4 N,

so the effective force driving the system is $(2.5\,T - 2354.4)$ N

Again, applying the equation $\Sigma F = ma$ to the entire system:

$2.5\,T - 2354.4 = 4440\,(0.1) \therefore T = 1119$ Nm.

The reader may confirm that in order to raise the lift at constant velocity, the torque needs to be 941.8 Nm.

If the lift with three people aboard needs to be accelerated downwards at 0.1 m/s², would it need to be driven or braked?

Conclusion

To solve problems in linear inertia, only one simple equation is required, namely the acceleration equation, $\Sigma F = ma$.

However, it is difficult to find examples that do not include the need to take into account the inertia of rotating components.

The preceding examples have been done assuming that the rotating parts in the mechanical linkages described, such as sheaves and drums, have negligible mass and do not experience friction.

In reality, all rotating parts have rotational inertia, due to their mass and dimensions. Such rotational inertia will definitely affect the acceleration of any system of which they are part.

Additionally, the frictional torque in the bearings of a rotating item will also affect its acceleration.

Once we have explored rotational inertia, we will be better equipped to deal more realistically with the inertia of systems that contain both linearly moving and rotating parts. This is done in the chapter on 'rotational inertia' in vol. 3 of this series.

Chapter 16

Linear Momentum

We distinguish between linear momentum (applied to objects in linear motion) and angular momentum (applied to rotating objects), which is not dealt with here, but in a subsequent chapter in volume 3 of this series. In the present chapter we will call linear momentum simply 'momentum'.

Definition of momentum

The ways in which transfers of momentum are used intentionally

The law of conservation of momentum

Collisions in one dimension

Definition of an impulse

The force during an impact: the reality and the approximation

Collisions in two dimensions

The steady-state transfer of momentum by a fluid stream

Definition of momentum

A moving object, for example a battering ram or an arrow, is capable of exerting a force on something else, by virtue of the fact that it has velocity, and it has mass. The greater that both of these quantities are, the greater the force it can exert. The momentum of an object is defined to be the product of its mass and its velocity.

Traditionally, momentum is given the symbol 'p' (lower case).

For an object of mass m, moving with velocity v, momentum **p = mv**

The units of this vector quantity are (units of mass)(units of velocity) = kg.m/s or N.s

The term 'momentum' came into use in the 1600s, around the time of Newton. However, Newton himself appears not to have used this term. He called what we now call momentum 'the quantity of motion' that a body possesses. So, what is the significance of this quantity?

One could think of the momentum of an object as an indication of how difficult it would be to bring that object to rest.

If either the mass or the velocity of an object is large, or if both are large, we are aware of the momentum it possesses and would treat that object with caution. We know intuitively not to stand in front of a moving vehicle or the stream of water emerging from a fire-hose nozzle. Rugby players will understand the challenge of tackling an opponent of considerable mass moving at high velocity.

High values of momentum may be observed in many situations, for example, when:

- An ocean liner approaches the dockside (very large mass, despite low velocity)
- A bullet of small mass moves at high velocity
- Water comes down a river in flood (large mass, considerable velocity)
- A runaway truck goes careering down a hill.

When one moving object collides with another, they exert forces on each other which result in changes of velocity for both of them. A change of velocity implies a change of momentum. We say that momentum is *transferred* from one colliding object to the other, as a result of a collision.

Transfers of momentum can occur with an unwanted destructive effect, such as when vehicles crash or boulders from rockfalls hit buildings. However, momentum transfers are frequently made to happen *intentionally.*

The ways in which transfers of momentum are used intentionally

Intentional transfers of momentum occur when, for example:

- Balls are struck by tennis racquets, golf clubs, boots or bats
- A hammer is used to knock a nail into wood
- A pile driver drives a pile into the ground
- A piece of sheet metal is forced into a die by a hydraulic press
- A mallet strikes the back of a chisel
- Bullets, arrows and other solid missiles strike targets
- Billiard balls are propelled to hit other billiard balls
- A battering ram is used to break down a door
- A jet of ultra high pressure water is used to cut steel
- A rocket is propelled by a stream of emerging exhaust gases
- The water in a river drives a waterwheel
- The wind is employed to operate a windmill or drive a sail

Among the examples listed above, some involve the transfer of momentum through a collision, while others illustrate instances of momentum transfer by a stream of fluid impinging on an object.

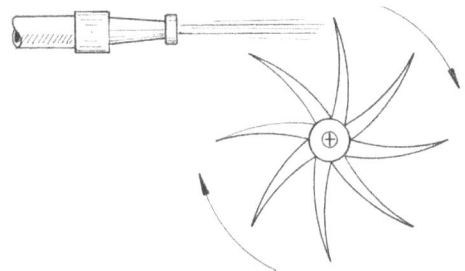

Having an understanding of momentum allows us to analyse the likely outcomes of collisions, and sometimes to work back from an observed outcome, for example to determine the probable speeds of crashed vehicles at the time they collided.

An understanding of momentum can enable us to determine the magnitudes of forces in certain situations, particularly in relation to momentum transfers by a fluid stream.

The law of conservation of momentum

It was emphasised in an earlier chapter that when analysing collisions, it is futile to try to do calculations based on the principle of conservation of energy, because too much energy is 'lost' (i.e. remains irretrievable) as a result of an impact. In order to analyse what velocity changes occur as the result of a collision, we have to look at the momentum transfer that occurs, because all the momentum of colliding objects *is* conserved.

The law of conservation of momentum states that for a given system of bodies, the total amount of momentum of the system remains unchanged, unless there is an input of momentum from *outside* the system.

This law is one of the basic pillars of the science of mechanics. In order to understand the derivation of this law, it is necessary to have a thoroughly clear grasp of two particular principles: Newton's 3rd law, and what the present author calls the zeroth law of mechanics. These two principles have both been explained in an earlier chapter on forces. Here's a brief recap to put them in context:

Newton's 3rd Law: *If two objects, A and B, have an interaction, the force that A exerts upon B is equal and opposite to the force that B exerts upon A.*

This law sounds deceptively simple. It is especially confusing that people have given Newton's 3rd law the misleading abbreviation: 'action equals reaction'. When you hear this phrase, you expect that when two objects come into contact, the *effect* of one object upon the other is going to be same.

However, it is not the *effects* that they have on one another that are equal and opposite. While it is true that the *forces* that they exert on one another are equal and opposite, the *effects* of these forces can be vastly different, depending on the masses and other properties of the two objects concerned.

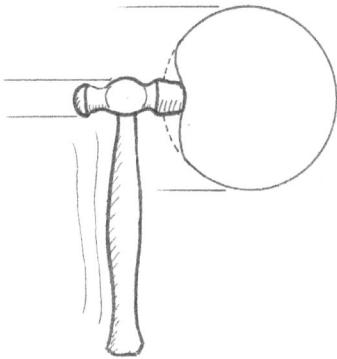

One might ask, for example, how does Newton's 3rd Law explain what happens when a flimsy inflated beach ball is hit with a heavy hammer? It is tempting to believe that the force that the hammer exerts on the ball is greater than that which the ball exerts on the hammer. Actually, neither is greater. These forces are *at all times* equal and opposite. To appreciate this, we have to go right back to:

The zeroth law of mechanics: *No force can even exist without an equal and opposing force being present.*

It is impossible to exert a force anywhere, without that force being opposed. You have got to have something to pull against, or something to push against, or else there *will be, and can be, no force*. For example, most people can easily exert a force of 20 N by pulling against a spring scale. However, just try to exert a force of 20 N, without pushing or pulling against something. It is impossible.

When most books on mechanics present Newton's 3rd law, what they do not add in so many words, is that the *effects* of the 'equal and opposite' forces upon the two opposing objects can be markedly different. Whatever the force is between the hammer and the beach ball, let us say, for the sake of argument, it reaches a peak of 50 N: both objects experience the same force, but that magnitude of force, acting for the short time that it does, will be sufficient to rocket the beach ball away, while it will hardly disturb the swing of the hammer.

Why momentum is conserved

Consider a set of objects (forming a 'system') that might interact or that might experience forces from outside the system. A system could comprise, for example:

- The marbles in a ring drawn on the ground
- The stars in a galaxy
- Two trains on the same track
- A bat and a ball
- Some people in a boat, including the boat.

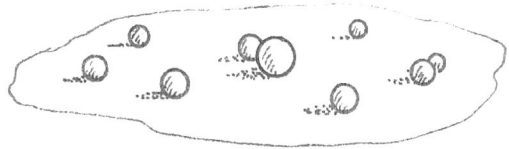

When we talk about a system of objects, we automatically imagine a boundary line that defines the border of the system. Anything outside this line we call external to the system. Within any such system the following logic applies:

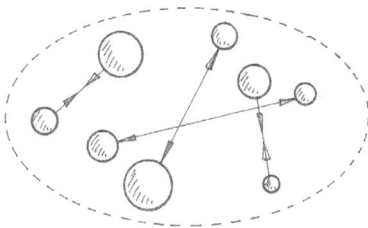

- If two objects within a system interact, they exert equal and opposite forces on one another for the duration of that interaction. Even if the forces they exert change in magnitude with time, at all times while they interact, the forces they exert are equal and opposite.

- If one of these objects subsequently interacts with a third object that is part of the system, then, in turn, the force that those two objects exert on one another will also be equal and opposite.

A 'system' of objects could also initially consist of one object, which subsequently divides up into smaller pieces, due to some internal process. Examples of this would be:

- If there is an explosion inside the object, ripping it apart (like a grenade)

- If the object was split to begin with, and the two pieces were held together by some means that would resist a compressed spring placed between them, with the potential to drive them apart. When the spring is released, it acts upon both surfaces with equal force.

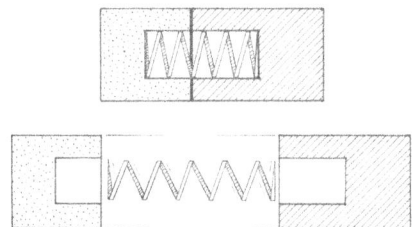

155

Once we define what makes up a system of objects, no matter what the interactions are *within* that system, all the forces that are exerted among the objects comprising the system occur in equal and opposite pairs. This means that, within the system, these forces cancel one another out.

For example, consider a system consisting of some sailors on a wooden boat with a deck. If two groups of sailors conduct a tug-of-war on the deck, no matter how hard they pull, in whatever direction they choose, they will not change the momentum of the boat.

However, if they were to pull on a rope attached to a fixed point on land, naturally they would change the momentum of the boat. The tension in the rope, in this case, constitutes a force that is *external* to the system.

An external force applied to a system of objects will definitely change the momentum of the system. External forces can also arise from:

- intrusion of one or more objects from outside the system,
- loss of objects from the system,
- streams of fluid entering or leaving it, or
- exposure of the system to a change in the magnetic or gravity field.

Derivation of the law of conservation of momentum

For any given object within a given system of objects, the rate of change of its momentum with time is expressed as the derivative dp/dt.

Since $p = mv$, $dp/dt = (m)(dv/dt) = ma = F$, where F is the total force acting on that object.

For a system of objects named respectively 1, 2 , 3…n, the total momentum

$$p_{tot} = p_1 + p_2 + p_3 + … p_n$$

(*bear in mind that this is a vector equation, not an arithmetic one. Each term represents a vector quantity, and the plus signs indicate vector addition, not arithmetic addition.*)

$$\therefore dp_{tot}/dt = F_1 + F_2 + F_3 + …F_n \dots\dots\dots\dots\dots\dots\dots(1)\ (also\ a\ vector\ equation)$$

where, for each object in the system, F represents the resultant of all the forces acting on that object. So, for example, F_1 = the vector sum of the external forces acting on object 1 *and* the internal forces acting on object 1.

156

Noting that the vector sum of all the *internal* forces among the objects comprising the system has to be zero, the vector sum of all the forces on the RHS of equation (1) amounts to the vector sum of all the *external* forces acting on the system.

$$\therefore\ dp_{tot}/dt\ =\ (F_1 + F_2 + F_3 + ...F_n)_{external}\ =\ \Sigma F_{external} \cdots\cdots\cdots\cdots\cdots\cdots\cdots\cdots\cdots(2)$$

Now, very importantly, *if* the sum of the external forces acting on a system is zero: then $dp_{tot}/dt = 0$, which means the rate of change of its momentum is zero, which implies that there is no change in momentum for the system as a whole.

In other words: in the absence of external forces, the amount of momentum contained in a system of objects remains constant.

This statement constitutes the law of the conservation of momentum, which has many applications. For example:

Collisions in one dimension

If two objects moving in the same straight line collide, their total momentum before the collision equals their total momentum after the collision.

This is expressed in the equation:

$$m_1 v_1 + m_2 v_2 = m_1 v_1' + m_2 v_2' \cdots\cdots\cdots\cdots(3)$$

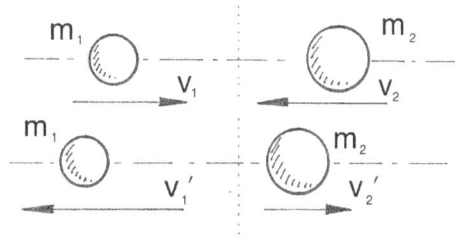

In equation (3) we have introduced the notation v_1' and v_2'. These are pronounced as 'v-one-prime', and 'v-two-prime'. We usually distinguish the velocities of objects *after* a collision by the addition of a 'prime' sign.

The commonest examples of controlled collisions in one dimension are found in the games of billiards, snooker and pool. The balls used in these games are made to a standard size and mass, from material that behaves elastically in the range of forces to which they are subjected in normal use.

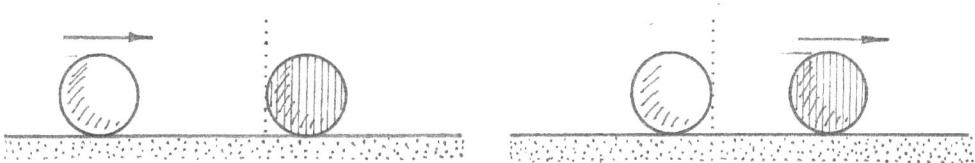

It is readily observed in practice is that, if one ball is made to hit another, moving in a direct path along the line joining their centres, the first ball remains standing at the position of impact, while the second ball moves off along that line with the same velocity that the first ball possessed before the impact.

In such a case, *all* the momentum of the first ball has been transferred to the second ball.

Suppose two or more balls are placed in a straight line, all touching the ones on either side of them. Now, if another ball is rolled along a path coinciding with the line joining their centres, to hit the ball at the start of the line, what happens? It is found that the ball at the far end of the line moves off, while all the others in-between remain stationary.

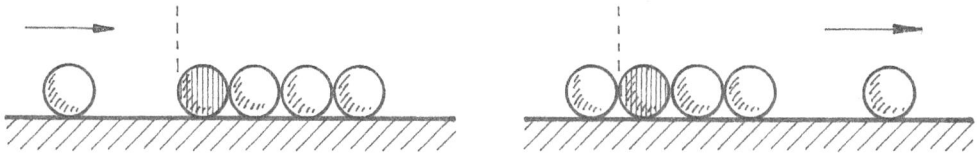

In this case, all the momentum of the first ball has been transferred to the *last* ball in the line. This happens so quickly that it appears instantaneous. The sight of this is fascinating, particularly as we intuitively imagine that all the momentum of the first ball ought to be shared out between the ones that were lined up, and that they would all move off with some reduced velocity. If the balls that were lined up were glued together, this is indeed what would happen.

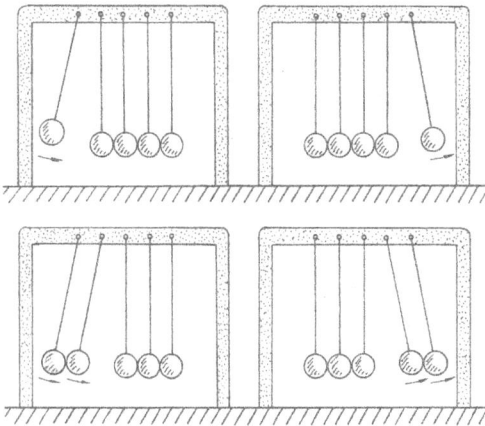

However, the transfer from the first ball to the last has not been direct. The original momentum has been passed from one ball to the next along the line, in the same manner as when one ball impacts upon another as illustrated previously.

The desk toy known as 'Newton's cradle' functions on the same principle. If five suspended balls are all touching, and you raise one ball at the end of the row and let it drop to strike the adjacent ball, only the ball at the far end swings up. The ones in-between remain stationary.

If you raise two of the balls at one end and let them fall together, then only the two balls at the far end will swing upwards. If you raise four balls together, and let them swing down, then the last four balls will swing upwards.

This toy demonstrates that the total amount of momentum of a system remains conserved.

Example

Suppose a direct collision occurs between two balls, A (2 kg) and B (7 kg), moving in the same straight line, with velocities respectively 6 m/s to the right and 3 m/s to the right.

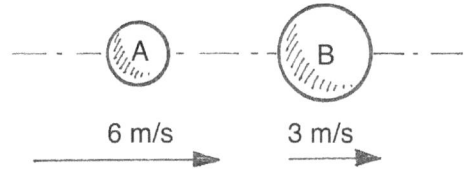

6 m/s   3 m/s

After the collision, the velocity of ball B is observed to be 4 m/s to the right.

Determine the velocity of ball A after the collision, and the percentage of the original kinetic energy lost in the collision.

Using equation (3): Initial momentum = final momentum $\therefore m_1v_1 + m_2v_2 = m_1v_1' + m_2v_2'$

Since this collision occurs in one dimension in this case, the vector equation may be treated as a linear algebraic equation. Consider the positive direction to be to the right.

$\therefore (2 \times 6) + (7 \times 3) = (2 v_1') + (7 \times 4)$ $\therefore v_1' = 2.5$ m/s to the right

Original KE = $\frac{1}{2}(2)(6)^2 + \frac{1}{2}(7)(3)^2 = 67.5$ J

Final KE = $\frac{1}{2}(2)(2.5)^2 + \frac{1}{2}(7)(4)^2 = 62.25$ J

The energy deficit is due to the production of sound and heat as a result of the impact.

$\therefore$ Amount unrecoverable (that is, 'lost') = 5.25 J, namely 7.778% of the original amount.

What happens to the momentum in a 'soft' collision, such as when a ball thumps into a lump of putty?

In such a case, no continuing movement is observed as a result of the collision. So, what does that say about the conservation of momentum? What happened to the momentum of the ball?

If the putty were floating in outer space at the moment of impact, it is easier to imagine that the ball and putty would move on from the impact with reduced velocity but unchanged momentum.

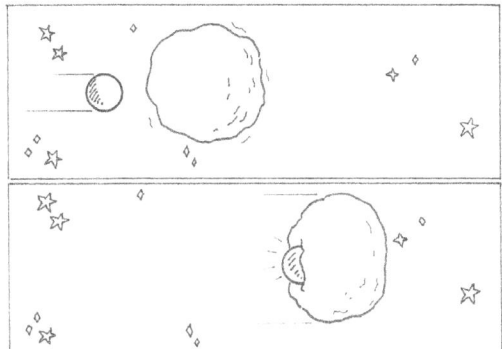

What makes it different if the collision happens on Earth is that the putty is restrained from free movement because it is grounded in some manner: either against a wall

or stuck to a table-top. This means that the momentum of the ball is ultimately transferred to the Earth, via the putty.

Now, the mass of the Earth is so large in comparison with the mass of the ball that the resulting change in the velocity of the Earth is not only infinitesimally small, but undetectable by us, since we measure all velocities relative to the surface of the Earth.

Also, for the ball to have been given the velocity it had in the first place (before the collision), it was necessary for there to have been an impulse imparted to the ball. (An 'impulse' is defined on p 167). The opposite impulse must have been experienced by the Earth, as something must have pushed backwards against the Earth, in order for the ball to have been launched forwards. This means that the Earth experienced first a backward impulse and then, at the moment of impact, a forward impulse of the same magnitude, which two impulses effectively cancel each other out, rendering the movement of the Earth unaffected. This has a direct parallel with the previous example of the sailors having a tug-of-war on the deck of a boat.

The application of the Law of conservation of momentum depends very much on how we define the system of objects that are involved in a momentum transfer. The sailors and their boat were defined as one system. The ball, the putty and the Earth are another.

It should be evident that exactly the same reasoning applies to a bullet that penetrates an earth bank behind a target.

Example

A horizontal air-bed rail (shown below) is a piece of equipment constructed specifically to demonstrate the principle of the conservation of momentum. Two 'sliders' rest on the rail. When the air flow is activated, the sliders are kept floating a small distance above the rail by air under pressure emanating from tiny holes in the rail. This allows the sliders to be moved horizontally along the rail with a minimum of friction.

Suppose that two purpose-built 'sliders' of accurately known mass are travelling on such a rail. Suppose also that they are fitted with a mechanical coupler which causes them to be locked together after colliding.

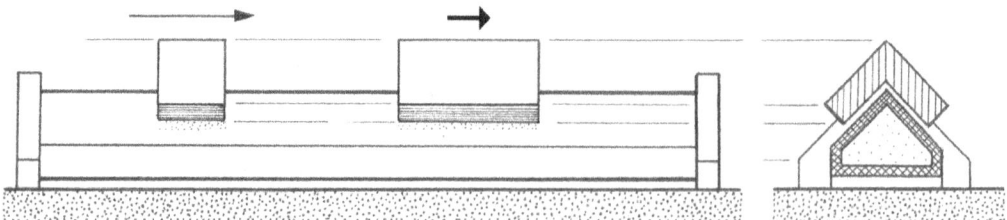

Slider 1 (2 kg) is moving at 7 m/s to the right when it catches up with and collides with slider 2 (4 kg) which is moving at 1 m/s to the right. Consider the positive direction to be to the right. Determine:

* the velocity of the linked sliders after impact.
* the percentage of the initial kinetic energy lost in this collision.
* the equivalent values, if the sliders were initially travelling *towards* one another.

Solution

The initial momentum of the system comprising these two mass-pieces:

$p_1 = m_1 v_1 + m_2 v_2 = (2 \times 7) + (4 \times 1) = 18$ N.s (*positive value implies: directed to the right.*)

Let the final velocity of the combined masses after the collision be v'.

The total momentum after the collision must be equal to that before the collision

$\therefore p_2 = (m_1 + m_2) v' = 18$ $\therefore (2 + 4)v' = 18$ $\therefore v' = 3$ m/s.

The loss of kinetic energy arising from this collision:

Initial KE $= \frac{1}{2}(2)(7)^2 + \frac{1}{2}(4)(1)^2 = 51$ J

Final KE $= \frac{1}{2}(6)(3)^2 = 27$ J $\therefore$ Loss $= 24$ J or 47% of the initial amount.

Now, look at the situation in which these two masses are moving *towards* each other with the same speeds as before.

Since the velocities are vectors, we have to take into account the signs denoting their directions.

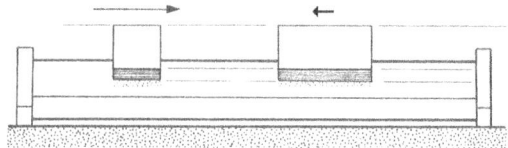

Initial momentum: $p_1 = m_1 v_1 + m_2 v_2 =$
$(2 \times 7) + (4 \times (-1)) = 10$ N.s (*again, to the right*)

Momentum after the collision: $p_2 = (m_1 + m_2) v' =$ initial momentum

$\therefore (2 + 4) v' = 10$ $\therefore v' = 1.667$ m/s (*still a positive value, hence to the right*).

As before: examine the loss of kinetic energy due to the collision:

Initial KE $= \frac{1}{2}(2)(7)^2 + \frac{1}{2}(4) (-1)^2 = 51$ J (*as before*)

Final KE $= \frac{1}{2}(6) (1.667)^2 = 8.333$ J, representing a loss of 42.67 J, which amounts to 83.66% of all the kinetic energy that was available before the collision.

In general, it will be found that when the objects are travelling *towards* one another before a collision, a greater proportion of their kinetic energy will be lost, than if they collided while travelling in the same direction.

It is even quite possible in some instances that *all* of the original kinetic energy will be lost. This would occur, for example, if the colliding objects had the same mass, and approached each other at the same speed relative to the ground, provided there was some way of keeping them together when they collide. If there was no means of keeping them together, they would bounce off one another, and retain some of the initial kinetic energy.

Exercises, set 1: one-dimensional collisions

Question 1

Two balls collide directly, along a straight line joining their centres, as shown.

Ball B rebounds to the right at 1 m/s after the collision. Determine the velocity of ball A after the collision, and the percentage of the original kinetic energy lost in the collision.

[1 m/s to the left; 97.35%]

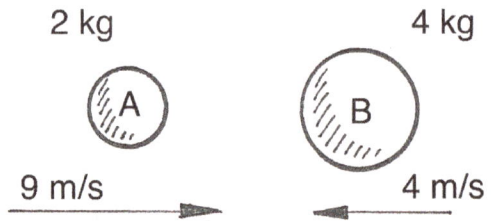

2 kg   4 kg

A   B

9 m/s   4 m/s

Question 2

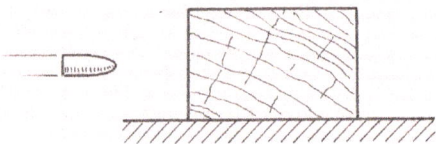

A bullet of mass 30 g, moving horizontally at 540 m/s, strikes and is imbedded in a 5 kg wooden block resting on a smooth horizontal table.

Determine the velocity of the block immediately after the collision. [3.221 m/s]

Question 3

A naval gun of mass 10 tonnes fires a shell of mass 100 kg horizontally with a muzzle velocity of 400 m/s. The recoil is resisted by a set of coil springs that must bring the gun to rest over a distance of 200 mm. Determine:

- The initial velocity of recoil of the gun. [4 m/s]
- The required combined stiffness of these springs. [4 × 10^6 N/m]
- The maximum force against the springs during deceleration. [800 kN]

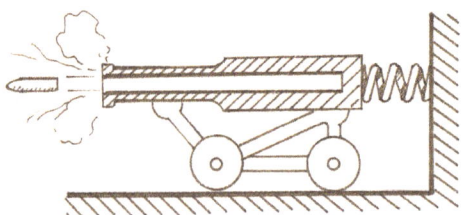

162

- The number of men that would be required to restrain this gun from recoiling, if they had to take the place of the grounded recoil-resisting springs, and bring the gun to rest over a distance of 200 mm. Suppose each man can push with a force of 125 N.   [6400 men]

Collisions preceded by or followed by other velocity changes

Stating that 'Initial momentum = final momentum' we saw that

$$m_1v_1 + m_2v_2 = m_1v_1' + m_2v_2' \dots\dots\dots\dots\dots\dots\dots\dots\dots(3)$$

Given the values of three of these velocities, we could solve for the remaining unknown velocity. Given only the values of the velocities *before* the collision, we could not solve for v_1' and v_2', unless we had another equation relating v_1' to v_2'.

In an earlier example, we ensured that v_1' would be equal to v_2' by introducing a mechanical means of keeping the colliding bodies together. If there is no means of keeping the objects together, we need additional information about the velocities of one or both of the objects after the collision.

Sometimes the additional information we need about the velocities of colliding objects can be provided by applying the law of the conservation of energy to what happens to the objects either before, or after the impact.

For example, suppose ball A , suspended by a hinged fine wire, is raised a height 'a' before allowing it to fall.

Ball A approaches ball B, which is similarly suspended, and collides with it at a velocity which can be determined by applying the conservation of energy to the down-swing of ball A.

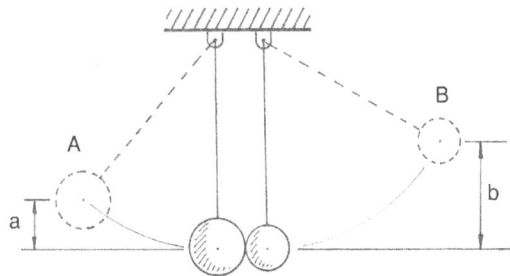

Suppose that ball B rose to height 'b' after the impact. One could apply the conservation of energy to the movement that occurs *after* the impact, to determine the velocity of ball B immediately after the collision.

These two velocities could then be used in equation (3) to determine the velocity with which ball A bounces back from the impact.

Exercises set 2: combining momentum and energy considerations to solve for unknowns in collisions

Question 1

A steel ball of mass 5 kg, attached to a hinged wire, is released when the wire is horizontal.

It swings down a vertical distance 'a' and hits an 8 kg block of hardwood suspended as shown.

After the collision, the block rises to a level that is height 'b' above its rest position, before swinging back again.

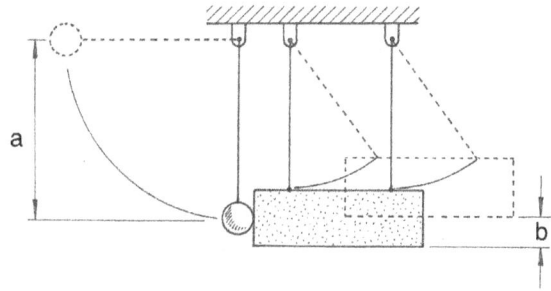

If a = 1.00 m and b = 0.220 m, determine:

  a. The velocity of the ball just before the collision   [4.429 m/s]
  b. The velocity of the block, just after the collision   [2.078 m/s]
  c. The velocity of the ball, just after the collision   [1.104 m/s]
  d. The percentage of the original kinetic energy of the ball that is lost in this collision   [58.57%]

Question 2

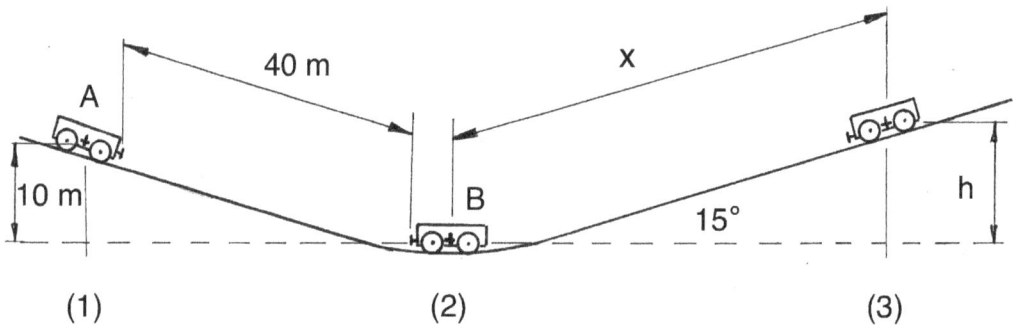

Two straight slopes are joined by a short smooth curve at location (2). Two identical strongly-built wagons, A and B, with wheels of negligible mass, and total mass 160 kg each, are placed in the locations (1) and (2). These wagons have sprung bumpers to minimise damage from impacts with one another.

Wagon A, which contains a load of 240 kg, is released from rest in location (1) and allowed to roll down the slope to collide with wagon B, which contains no load, and is initially stationary. Both wagons experience rolling resistance of 50 N.

Assuming that 90% of the initial kinetic energy is available after impact, (a randomly chosen figure, possibly justified by the fact that the bumpers have springs) and ignoring the effects of air resistance, determine:

•   The distance 'x' that wagon B rolls up the subsequent slope before slowing to a momentary stop at location (3).   [54.32 m]

•   The distance that wagon A rolls back up the slope towards location (1) after the collision, before slowing to a stop.   [8.374 m]

164

The ballistic pendulum:
a practical application that combines momentum and energy considerations

The first exercise in the preceding set illustrates the operating principle of the ballistic pendulum, invented by English military engineer Benjamin Robins in 1742. This device enables the determination of the speed of a projectile such as a musket shot or cannon ball, without the need for measuring time. Nowadays sophisticated equipment exists to measure time and speed directly. However, until Robins' invention, no-one had any way of knowing what the speed of a projectile was.

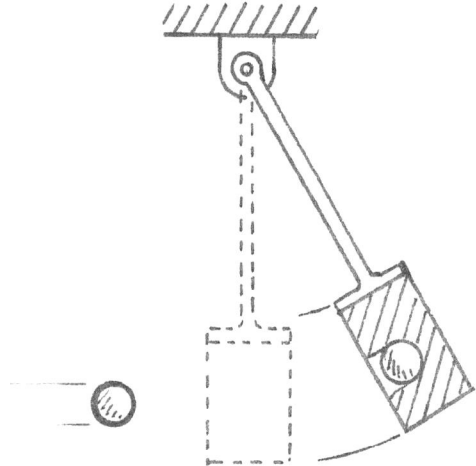

The ballistic pendulum works like this: A projectile is fired horizontally into a suspended pendulum block, and stays imbedded in the block. The block swings up a certain amount after the impact.

From the height of the swing, using the principle of the conservation of energy, the velocity of the block (with the projectile imbedded in it) directly after the impact can be determined. This velocity can then be used in the equation deriving from the law of conservation of momentum, to determine the projectile velocity before impact.

Example

A bullet of mass 40 g is fired into a ballistic pendulum suspended on hinged wires of length 1 m. The block that absorbs the bullet has mass 10 kg, and rises to a height of 112 mm before swinging back again.

Determine the impact speed of the bullet.

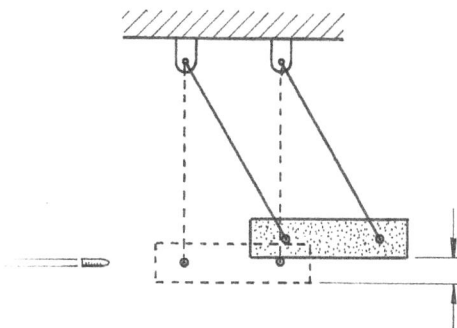

At the top of its swing, the block has potential energy:

$PE = mgh = 10.04(9.81)(0.112) = 11.031$ J

This amount of PE can be equated to the KE of the block at the start of its swing:

$KE = \frac{1}{2}m(v')^2 = \frac{1}{2}(10.04)(v')^2 = 11.031$
$\therefore v' = 1.4824$ m/s

165

Now apply the law of conservation of momentum to the collision:

Initial momentum of the block is zero. If the bullet speed is v_1 then $m_1 v_1 + 0 = (m_1 + m_2)v'$

$\therefore 0.04 v_1 = 10.04(1.4824) \quad \therefore v_1 = 372.1$ m/s

Initial KE $= \frac{1}{2}(0.04) (372.1)^2 = 2768.8$ J

KE after impact was 11.03 J, hence 99.60% of the original KE was lost.

Most of this energy goes into deformation of material, namely into work done by the bullet in driving a hole into the block, and work done to deform the bullet. There is also some loss due to the generation of heat and the production of sound at the impact.

Exercises set 3: ballistic pendulums

Question 1

An air-gun fires a 50 gram dart horizontally. The dart penetrates and remains imbedded in the surface of a ballistic pendulum suspended from four light wires of length 2 m. The mass of the pendulum block is 2 kg exactly. After impact, the block rises to the point where the wires are at 18° to the vertical, before swinging back. Determine the velocity of the dart, and the percentage of the KE of the dart that is lost in the impact. [56.82 m/s; 97.56%]

Question 2

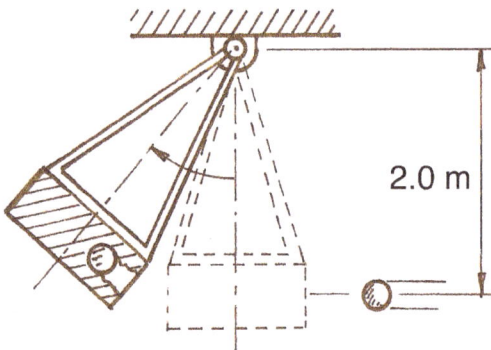

2.0 m

A ballistic pendulum consists of a 200 kg block of wood firmly attached to a steel frame that is hinged to a grounded support.

A 5 kg cannon ball is fired horizontally into the centre of one face of the wooden block. The ball remains imbedded in the wood, and the frame swings up to make an angle of 42° with the vertical before swinging back down.

Ignoring the mass of the steel frame, air resistance and any frictional resistance at the hinge, determine:

- the incident velocity of the cannon ball. [130.2 m/s] and
- the percentage of the original KE that is lost in the impact. [97.56%]

Question 3 *(suitable for discussion or for a design-and-build project: answers deliberately not provided)*

Suppose you have to design and construct a ballistic pendulum that is capable of determining the exact speed of a small cannon-ball (cast iron, diameter approximately 60 mm) whose speed is estimated to be possibly as much as 300 m/s. The four suspension rods of the ballistic pendulum, though strong enough for the shock of impact, may be considered to have negligible weight. They are each 1.5 m long. Answer the following questions, giving your reasoning, based on what you have learnt from the preceding example and exercise:

a. As a first approximation, what should be the mass of the pendulum block?

b. How would you design the block to ensure that the cannon-ball remains imbedded and does not bounce off it, or smash it to pieces?

c. How would you measure the height to which the block rises at the top of its swing? (if no electronic instruments are available.)

Definition of an impulse

A change in momentum, such as occurs during collisions, is brought about by the action of a force acting *over a period of time*. During a collision, this period of time is usually very short, but is nevertheless finite. In other circumstances a force may be applied over a longer period. Also, irrespective of the duration of the period, the force could either be of constant value, or such that it rises from zero to a peak value and back to zero.

Various examples of forces that result in a change of momentum:

- A wagon freewheeling with constant velocity down a straight gentle slope encounters a part of the slope that is steeper, before the slope reverts to its original incline.

- A train proceeding along a level track has the brakes applied for half a minute, to make it slow down.

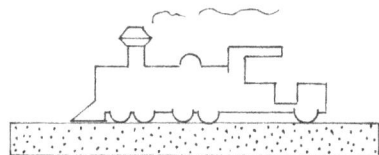

- A cricket ball moving to the left is struck by a bat moving to the right and accelerates rapidly to the right.

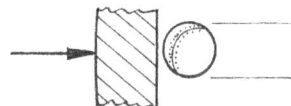

The force vs. time curves for the period of momentum change in these three examples would look roughly as follows, respectively:

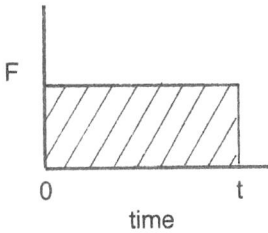

The wagon accelerating   **The train braking**   **The bat and ball**

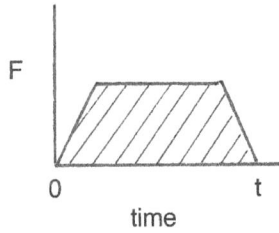

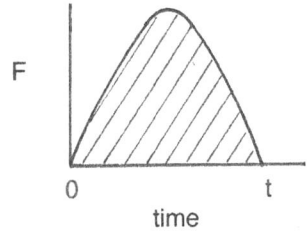

The object undergoing the change in momentum experiences what is called an *impulse*, namely that which is *responsible* for bringing about the change in its momentum. The impulse will be given the symbol j.

If the force remains *constant* for the duration of its action, as in the case of the wagon in the first diagram, the magnitude of the impulse of that force: $j = Ft$, and is represented by the area under the graph. In the other two cases, in which the force varies with time, the impulse is also represented by the area under the graph of force vs. time for the duration of the action of the force.

The impulse acting on an object is defined as the integral of the force function with respect to time, for the period of the momentum change.

$j = \int F(t)\, dt$ between $t = 0$ and $t = t$

An impulse is a vector quantity, with the same units as for momentum, namely: [units of force × units of time], that is: [kg.m/s] or [Ns].

The value of the impulse, when the force causing a momentum change is constant, or justifiably assumed to be constant

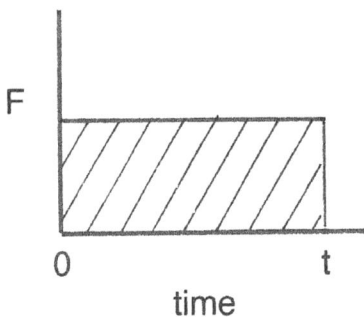

If an object of mass m is originally moving at velocity v, and has its velocity changed to v' after experiencing a force F, of constant value, for a limited time, t, then the change in momentum:

$$p_2 - p_1 = m(v' - v) \quad\dots\dots\dots\dots\dots\dots(4)$$

Also, the acceleration experienced by the object is given by the rate of change of its velocity with time: $a = (v' - v)/t$

However, the acceleration is also defined by $a = F/m$

$$\therefore (v' - v)/t = F/m \quad \therefore m(v' - v) = Ft \qquad \therefore p_2 - p_1 = Ft \dots\dots\dots\dots\dots\dots\dots\dots\dots(5)$$

Which illustrates the fact that the impulse acting on an object is equal to the change in momentum of that object.

In a collision, the impulses that the two colliding objects exert on one another are equal and opposite, since the forces that they exert on one another are at all times equal and opposite, and the time for which they are in contact has to be the same for both objects.

Since impulse $j = Ft = m(v' - v)$ for any one of the objects, if we know its velocities before and after the impact, and we are able to measure the time 't', we can deduce an *average* value for F, based on the assumption that F was constant during the impact.

Example

A train of mass 200 tonnes is moving along a level track at 10 m/s. The brakes are applied for a period of 5 seconds, reducing the velocity of the train to 6 m/s. Determine: the average value of the braking force, and the magnitude of the impulse.

$$Ft = m(v' - v) \quad \therefore F (5) = 200\ 000(10 - 6) \quad \therefore F = 160\ 000\ N = 160\ kN$$

$$j = Ft = 160\ 000(5) = 800\ 000\ \text{N.s or } 800\ \text{kN.s}$$

In the above example, the impulse was brought about by a moderately large force acting for a significant amount of time. In other situations, an impulse can be brought about by a very large force acting for a very short time.

In the example of the bat hitting the cricket ball, the significant change in the momentum of the cricket ball that is observed in practice indicates that the impulse must have been considerable. Since the time for which they were in contact was extremely short, we may conclude that the force between bat and ball had to have been very large.

A very large force acting for a very short time is called an 'impulsive force'.

Impulsive forces in perspective with other forces that surround a collision

Impulsive forces typically arise in situations such as a hammer blow, an item falling onto a surface from a significant height, explosions, and a bat striking a ball.

Usually the magnitude of an impulsive force is so large in comparison with all the other forces acting on the object at the time, that it is justifiable to consider these other forces to be negligible for the duration of the impact.

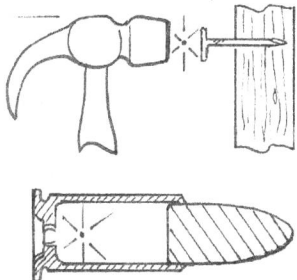

Such other forces could include, for example:

- The weight of the object being struck,
- Friction between the object and a surface with which it is in contact,
- The immediate initial force exerted by the springs in a recoil mechanism,
- Air drag on the object,
- Rolling resistance experienced by a wheeled object resting on a surface at the time of the collision, and
- The resistance to penetration offered by material, such as wood resisting the entry of a nail or by earth resisting the penetration of a pile.

Surrounding forces like these need to be taken into account up to the point of impact, and immediately following it, but since they may be assumed to be very much smaller than the impulsive force that acts *during* the impact, they may be ignored for the duration of the impact. Ignoring these forces is not a cop-out to obtain simpler calculations: you will see from the magnitudes of the impulsive forces in the following examples that they are orders of magnitude larger than all the other forces involved at the time of impact.

Example: Analysing the impact of a hammer driving a nail into wood

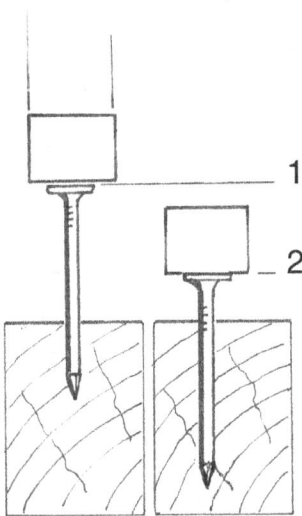

A hammer with mass 4 kg strikes a 50 g nail from above, driving it 20 mm into a block of wood. The speed of the hammer just before striking the nail is 5 m/s. Determine:

• The speed with which the hammer and nail together start moving into the wood.

• The average resisting force offered by the wood to oppose the entry of the nail, and

• How long it takes the nail to penetrate the wood this far, from this single blow.

Solution:

Initial momentum of hammer = mv = 4 kg × 5 m/s = 20 kg.m/s (or 20 N.s)

During the impact we will ignore the resistance to penetration offered by the wood, as this force is likely to be far smaller than the force of the impact.

Momentum after impact = (4 + 0.05) v', where v' is the velocity of the hammer and nail moving together after impact.

By the conservation of momentum, 20 = (4.05) v', ∴ v' = 4.938 m/s

After the impact, the nail moves steadily into the wood, doing work against the resisting force exerted by the wood. This motion is without impact and can thus be analysed using the principle of the conservation of energy.

In position 1, immediately after impact:

$KE_1 = \frac{1}{2}mv^2 = \frac{1}{2}(4.05)(4.938)^2 = 49.38$ J

Considering position 2 as the reference level for zero PE:

$PE_1 = mgh = 4.05(9.81)(0.02) = 0.795$ J

$\therefore$ Total available energy $= 49.38 + 0.795 = 50.18$ J

In position 2, when the nail comes to rest: $KE_2 = 0$ and $PE_2 = 0$

So all the energy it had in position 1 has gone into work done against the resistance offered to penetration by the wood, less some minor amount lost to heat and sound, which we will (reasonably) assume to be 20% of this quantity.

Mechanical work done $= Fd$, where $F =$ the average force exerted by the wood.

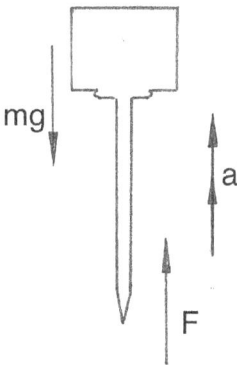

$\therefore$ 80% of 50.18 J $= 40.14$ J $= F(0.02)$ $\therefore F = 2007$ N

Consider the hammer and nail to behave as one object, decelerating as it moves into the wood. Let the upward direction be positive.

$\Sigma F = ma$ $\therefore 2007 - (4.05)(9.81) = 4.05a$ $\therefore a = 485.8$ m/s²

From $v = u + at$, $t = (v - u)/a$, $\therefore t = (0 - (-4.938) \div 485.8$

$= 0.0102$ seconds, namely 10.2 milliseconds.

Exercises set 4: Impulsive forces

Question 1

A pile-driver of mass 5 tonnes falls from a height of 3 m onto a pile of mass 1 tonne. If the pile is driven into the ground for 80 mm, what is the average resisting force exerted by the ground? [1592 kN]

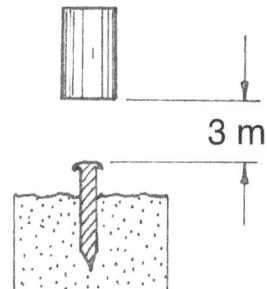

3 m

Question 2

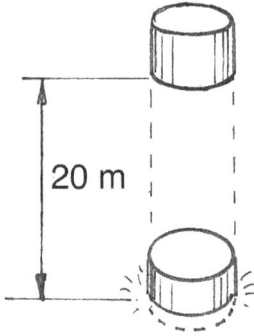

A solid disc of mass 18 tonnes and diameter 1.2 m is dropped onto the ground from a height of 20 m. The disc penetrates the ground by 100mm.

Determine the average resisting force offered by the ground, given that the density of the un-compacted soil is 1800 kg/m³.

Assume that a disc-shaped volume of this soil is pushed downwards a distance of 100 mm by the impact, and that 20% of the energy lost in the impact goes into generating sound and shock waves. [1.363 MN]

20 m

Question 3

A hammer of mass 1 kg, moving at 6 m/s, drives a 30 g nail horizontally into a well-supported piece of wood, to a depth of 25 mm. Determine:

a. The common velocity of the hammer and nail immediately after the impact [5.82 m/s]
b. The percentage of the hammer's kinetic energy that is 'lost' in the impact [2.91%]
c. The time for which the nail was moving [0.0086 sec]
d. The force with which the wood resisted the entry of the nail, assuming that force to have been constant. [699 N]

Question 4

A block of wood, of mass 5 kg exactly, is resting on a horizontal floor. The coefficient of kinetic friction between the block and the floor is 0.4. The block is struck horizontally by a bullet of mass 40 g moving at 350 m/s. The bullet remains imbedded in the block of wood. Determine:

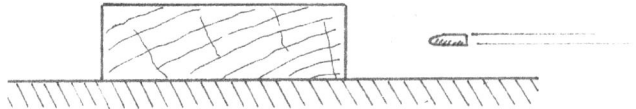

- The velocity with which the block and bullet together begin moving after the collision
- The distance which the block moves along the table.
- The ratio of the amount of kinetic energy lost during impact to the energy lost through subsequent friction with the floor. [2.778 m/s; 983 mm; 125:1]

Question 5

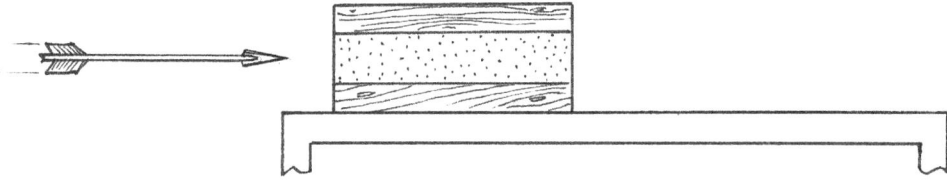

An arrow of mass 50 g is fired horizontally from short range into a target consisting of a slab of expanded polystyrene glued between two slabs of wood. The mass of the target is 4.0 kg. The arrow remains embedded in the polystyrene, and the whole assembly slides a distance of 86 mm along the steel table on which the target was resting. The coefficient of kinetic friction between the wood and the table is 0.3. Afterwards it is determined that the arrow penetrated the polystyrene by 56 mm. Determine:

- The velocity of the target immediately after the impact. [1.030 m/s]
- The velocity of the arrow immediately before striking the target. [83.43 m/s]
- The kinetic energy of the arrow before striking the target. [174.0 J]
- The percentage of this kinetic energy that was lost in the collision. [98.77%]
- The amount of work done for the arrow to penetrate the polystyrene, assuming that this constituted 80% of the energy loss (implying that an estimated 20% was lost to sound and heat) . [137.5 J]
- The average resisting force offered by the polystyrene. [2455 N]

The value of an impulsive force during an impact: the reality and the approximation

It is impossible to determine the true value of an impulsive force by calculation.

We saw that in the simple case where the force producing a change in momentum had a constant magnitude, the impulse $j = Ft$.

However, for most situations involving impacts, it is completely unrealistic to assume that an impulsive force would have constant magnitude. Most impulsive forces are likely to vary with time during an impact, rising from zero to a maximum and back to zero in patterns plausibly like those as shown below, though the exact patterns will be difficult to determine experimentally:

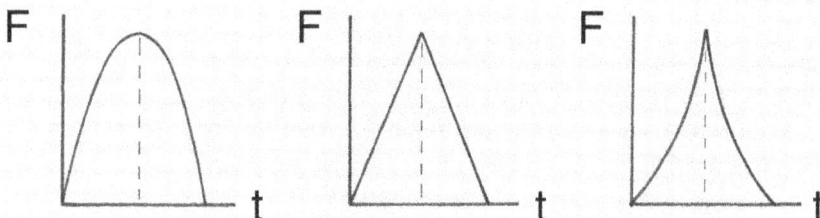

In most collisions, the time duration of the impact is usually so short (sometimes fractions of a millisecond) that it is *extremely difficult* to measure that time reliably, even with the most sophisticated equipment.

To understand how and why the impact force varies during a collision we have to be able to picture the way in which the materials of the colliding objects are affected by the impact.

No real objects are perfectly rigid. All objects that collide *will* deform to some extent during an impact. The extent to which they deform, and whether or not the deformation is permanent, greatly affect the outcome of the collision.

Collisions in which objects regain their original shape *entirely* after the collision are described as 'perfectly elastic'. Such collisions are usually encountered only in specific circumstances, such as with billiard balls or steel ball-bearings colliding, and at low velocities.

At the other end of the scale are those collisions in which both objects suffer extensive permanent deformation as a result of the collision. These are known as as 'perfectly plastic' collisions. For example: two lumps of putty that collide head-on at high speed.

Most real collisions are partially elastic and partially plastic.

Consider an elastic collision between a bat and a ball.

In the extremely short time for which they are in contact, the following stages occur:

1. Contact begins. The material of the ball that is touching the bat will begin compressing, even while the material at the far side of the ball is still moving towards the bat. This has the effect of temporarily flattening the ball. Similarly with the bat. Although the bat is larger, more massive, and made of different material, that part of the bat in contact with the ball also experiences compression.

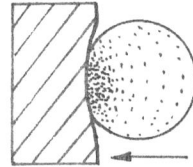

2. The force between the bat and ball keeps increasing until all the material of the ball has been decelerated to an instantaneous velocity of zero (relative to the bat), after which the elastic nature of the ball begins to restore its original shape. The elastic nature of the material of the bat also begins restoring the shape of the surface of the bat.

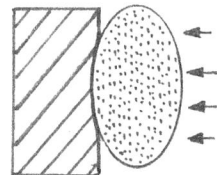

3. The force of the impact starts to diminish as the two begin to return to their original shapes. There is increasingly less compression, therefore increasingly less force between them, until this force becomes zero at the point at which the ball leaves the surface of the bat.

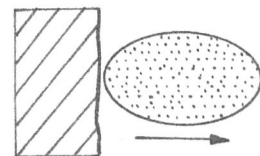

It must be fairly obvious that it would be a gross assumption to regard the force during any impact as constant. Clearly, the peak force between the colliding objects will always, in reality, be much greater than the average value.

A graph of force vs. time for the impact between ball and bat would probably look something like the graph of force vs. time for a solid rubber ball bouncing off a stationary flat surface, shown in the first of the three diagrams below. (Research has claimed that such a graph resembles part of a sine wave.)

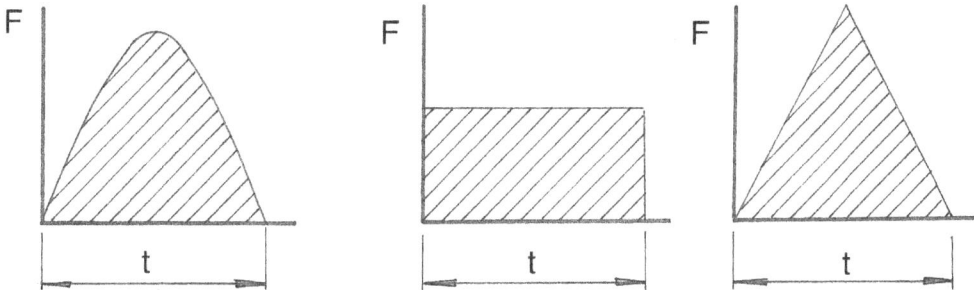

F   F   F

t   t   t

Since we don't know the exact shape of the force vs. time curve, we could *roughly* approximate the shape of the curve to a triangle. If we do this, comparing the 2nd and 3rd diagrams above, it appears that for there to be the same amount of area under the graph to represent the impulse, the peak force will be *at least twice* the average value of force over the duration of the impact.

The reasoning given above, describing the likely progression of events during the collision of a bat and a ball, applies to a near-elastic collision. Similar reasoning would also apply to collisions that are more plastic. However, in such cases the peak force is likely to be much smaller than for elastic collisions. This would be so because the relatively plastic material deforms, yielding rather than resisting the applied force.

Can we predict how much energy will be lost in a collision?

In an earlier section of this chapter, we said that it is very difficult to predict the fraction of the initial kinetic energy that will be lost in a collision. This is the case in most collisions, because usually there are so many variables involved, namely:

- There is an infinite assortment of shapes of objects that might collide, and we can't put a number to the degree of elasticity/plasticity exhibited by a given material, without also specifying the shape of the object that is involved in the collision,

- We can't make allowance for the large variety of combinations of materials from which the colliding objects might be made, and

- We can't tell, without experimentation, what the effect of different collision velocities will be. Some objects like billiard balls will behave elastically at low speeds, but shatter at high speeds.

We can, however, develop an index that predicts percentage energy loss for regularly-shaped objects of uniform material, all of the same shape, such as solid or hollow spheres colliding with others like them, or hitting a flat wall.

An index that fulfils this function is called the coefficient of restitution.

The coefficient of restitution, e

In everyday terms, this coefficient is a measure of how much *bounce* there is in a collision.

The coefficient of restitution is a number that represents the ratio of the relative velocity of the colliding objects *after* a collision to their relative velocity *before* that collision.

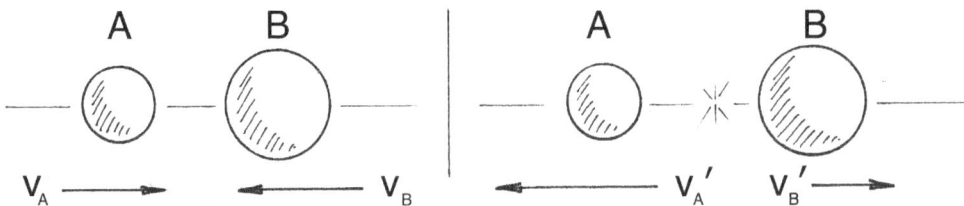

For spheres A and B that collide, with initial velocities v_A and v_B, and final velocities v_A' and v_B', the coefficient of restitution is defined as:

$$e = |(v_A' - v_B') \div (v_A - v_B)| \dots\dots\dots\dots\dots\dots\dots\dots\dots\dots(6)$$

In a perfectly elastic collision, the objects bounce back with the same relative velocities as they had before impact. The value of e = 1. The objects also possess the same total kinetic energy that they had before the collision.

In a perfectly *inelastic* collision, the objects stick together with no bounce at all, implying the relative velocity of the two objects after impact is zero, so e = 0. Also, all of the initial KE is lost.

The collisions between pairs of spheres made from most common materials exhibit an e-value somewhere between 0 and 1.

Example

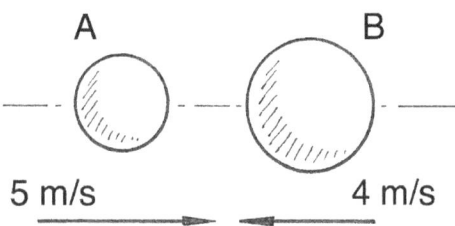

Consider a direct collision between two spheres of the same material, moving towards one another. Sphere A (3 kg) is moving to the right with velocity 5 m/s before the collision. Sphere B (6 kg) is moving to the left at 4 m/s.

If sphere A is observed to move to the left at 5 m/s after the collision, determine:

- The velocity of sphere B after the collision
- The percentage of kinetic energy that is lost in this collision, and
- The value of the coefficient of restitution for these spheres.

Solution: Let the velocities after the collision be v_A' and v_B'.

By the conservation of momentum, $((3 \times 5) + (6 \times (-4)) = (3 \times (-5)) + 6\,v_B'$

$\therefore v_B' = 1.0$ m/s to the right

Total KE before the collision: $\tfrac{1}{2}(3)5^2 + \tfrac{1}{2}(6)(-4)^2 = 85.5$ J

Total KE after the collision: $\tfrac{1}{2}(3)(-5)^2 + \tfrac{1}{2}(6)(1)^2 = 40.5$ J

Amount lost = 45.0 J $\therefore$ percentage lost = $(45.0/85.5) \times 100 = 52.63$ %

The coefficient of restitution, by equation (6) above:

$e = |\,(v_A' - v_B') \div (v_A - v_B)\,| = |(-5 - 1) \div (5 - (-4))\,| = 0.6667$

Where the coefficient of restitution is used

The value of 'e' cannot be specified for a material or an object on its own. This coefficient can only be determined for a *given pair* of colliding objects, either two spheres or a sphere against a flat surface. For instance, a golf ball falling onto a concrete floor will bounce to a certain height, but the bounce height will be different if the same ball is allowed to fall onto a carpeted floor.

Most spherical sports balls, including golf balls, tennis balls, basket balls and table tennis balls, have their coefficient of restitution regulated to an industry standard. In order to be acceptable, when the balls are dropped from a given height onto a standard surface, the height of the bounce must be within specified limits.

For the case in which a moving sphere, A, collides with a flat immovable object, B, such as a wall or floor: From equation (6) above, we see that, since both the initial and final velocities of B will be zero, the equation reduces to:

$e = |v_A' \div v_A|$

The velocity of an object is proportional to the square root of the amount of KE possessed by that object, so, if the collision is one of the sort where an object is dropped from a height onto a floor:

$e = \sqrt{(KE_{after} \div KE_{before})} = \sqrt{(PE_{after} \div PE_{before})} = \sqrt{(\text{bounce height} \div \text{drop height})}$

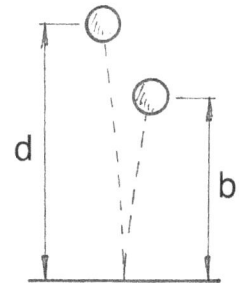

Example

A newly manufactured tennis ball is dropped from a height of 1000 mm onto a hard concrete floor, and bounces up to a height of 715 mm before falling back again. Determine the coefficient of restitution for this ball on this floor.

e = √(bounce height ÷ drop height) = √(715 ÷ 1000) = 0.846

Exercise

If air resistance was not a factor, how high would you expect the ball in the above example to bounce on the same floor, when dropped from a height of 3000 mm? [2145 mm]

Note: the value of 'e' exhibited by a ball hitting a flat surface will depend on the velocity of impact, which is in turn influenced by air resistance. For example, if you dropped the same tennis ball from a height of 100 m, by the time of impact it will have slowed down significantly from the speed it would have attained in a vacuum. It will also slow down significantly on its way up, after the bounce, so it cannot be expected to bounce to a height of 71.5 m.

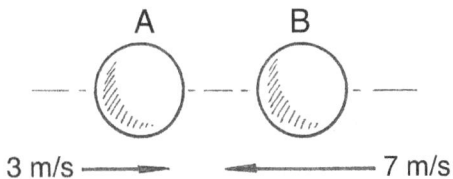

Exercise: Two balls of equal mass collide in a direct line, ball A moving at 3 m/s to the right and ball B moving towards it at 7 m/s to the left. It has previously been established that the coefficient of restitution between these two balls is 0.9. Determine their respective velocities after the collision.

[v_A' = 6.415 m/s to the left; v_B' = 2.415 m/s to the right]

The limited nature of situations in which the coefficient of restitution can be applied

With the exception of some billiard ball collisions, Newton's cradle and situations that can be set up under controlled conditions in a laboratory, most collisions do *not* involve one perfect sphere hitting another in a direct line, or hitting a flat wall or floor at right angles. An index like the coefficient of restitution is therefore of very limited value in analysing general and more complicated collisions, for the following reasons:

By definition, the coefficient of restitution can *only* be applied to one-dimensional *direct* collisions. By 'direct' is meant that the velocity vectors are aligned with the line joining the centres of mass of the colliding objects. So this coefficient cannot be applied to glancing collisions.

Also, while equation (6) does not make allowance for whether the velocities are high or low, certain experiments have found that the value of 'e' departs from the value predicted by equation (6) at very *low* velocities.

At significantly high velocities, the colliding objects could suffer more permanent deformation than they would at low velocities, resulting in a more plastic collision than that at low velocities. This would result in a different value for 'e'.

This coefficient therefore appears useful only as a quality control measure for sports balls.

Collisions in two dimensions

We know that the equation stemming from the principle of the conservation of momentum, namely:

$$m_1 v_1 + m_2 v_2 = m_1 v_1' + m_2 v_2' \quad \dots\dots\dots\dots\dots\dots\dots\dots\dots\dots\dots\dots\dots(3)$$

is a vector equation. When dealing with one-dimensional problems, we treated it as a linear algebraic equation. However, we cannot do this when applying the equation to 2-dimensional (or 3-dimensional) situations.

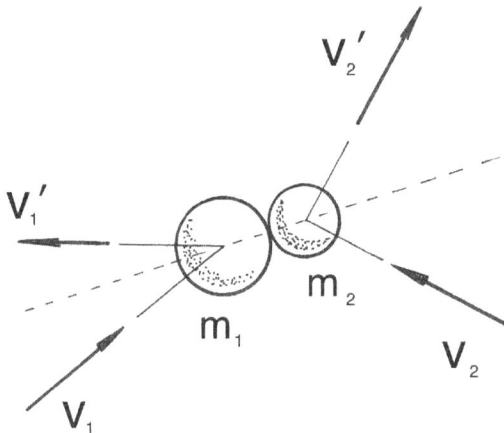

For a two-dimensional collision, we have to resolve the momentum vectors of the two colliding objects into components, respectively in the x- and y-directions.

We can then set up separate equations reflecting that momentum must be conserved in *each* of these principal directions.

It is readily apparent that in a two-dimensional problem there will be many more unknowns than in a one-dimensional problem.

Applying the conservation of momentum equation:

In the x-direction: $m_1 v_{1x} + m_2 v_{2x} = m_1 v_{1x}' + m_2 v_{2x}' \quad \dots\dots\dots\dots\dots\dots\dots(7)$

In the y-direction: $m_1 v_{1y} + m_2 v_{2y} = m_1 v_{1y}' + m_2 v_{2y}' \quad \dots\dots\dots\dots\dots\dots\dots(8)$

Appearing in these two equations are altogether *eight* velocity components. Since two equations enable us to solve for only two unknowns, we can only solve these equations if we have additional information that enables us to reduce the number of unknowns to two.

The reduction in the number of unknowns can come from combinations of the following circumstances:

- If the initial velocity of one of the colliding objects is zero,
- If the velocity of one of the objects after impact is known through some measurement, or can be determined by applying the conservation of energy to its subsequent movement,
- If the directions in which both objects move after the collision are defined. This condition would be satisfied realistically if there was some constraint applied to the motions, such as that they remain stuck together after impact, rendering their velocities after the collision to be identical.
- If the collision can be considered perfectly elastic, in which case no kinetic energy would be lost, allowing an energy accounting equation to be set up.

Example

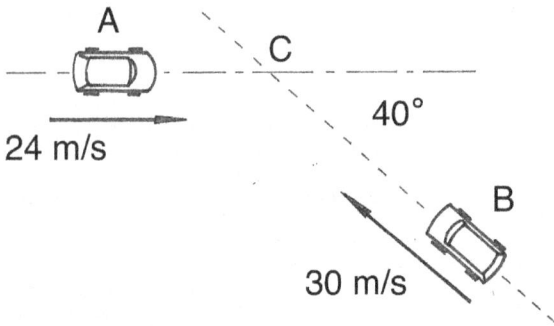

Two cars driving on a flat horizontal surface collide at point C, with the velocities shown.

Car A, travelling at 24 m/s has mass 1200 kg, and car B, travelling at 30 m/s has mass 1500 kg.

As a result of the impact, the cars become entangled and move together until friction with the road brings them to rest.

Determine the magnitude and direction of their common velocity immediately after impact. Consider the cars as mere masses, ignoring the tendency of a car to move in the direction that the wheels are pointing.

Solution:

Initial momentum in the x-direction: $1200(24) - 1500(30 \cos 40°) = -5672$ N.s

Initial momentum in the y-direction: $0 + 1500(30 \sin 40°) = 28925$ N.s

If the velocity after impact is v',

the momentum after impact is $(1200 + 1500) v'$

∴ in the x-direction:

$2700 v'_x = -5672$ ∴ $v'_x = -2.101$ m/s

in the y-direction:

$2700 v'_y = 28925$ ∴ $v'_y = 10.917$ m/s

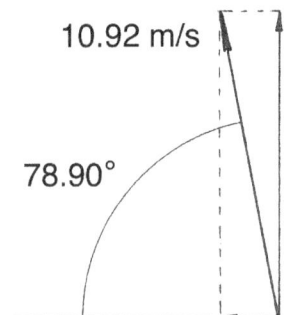

180

Re-combining these components: v' =10.92 m/s (or 39.3 km/h); W 78.90° N

Two-dimensional billiard ball collisions

To analyse a 2-D collision between billiard balls, we are justified in making the following assumptions:

- The masses of the two colliding balls are the same (which is not strictly true, as the cue ball is manufactured to have slightly greater mass than the object balls). If we don't want to assume the masses are equal, we can simply use the actual values. They are in the region of 170 grams, though the exact masses differ according to the equipment specifications for different types of game: whether billiards, snooker, pool or carom.

- The collision is perfectly elastic (such collisions are indeed as close as they can be to perfectly elastic, for all practical purposes).

- The surface sliding friction between balls is negligible, so there are no tangential forces at the point of contact (We have to assume this, in order to proceed with the limited equations we have, although it is likely that the friction coefficient between balls is significant.)

- No spin is involved, so there is no kinetic energy of rotation to consider. (Acknowledging that some players have the skills to impart spin deliberately.)

- The object ball is stationary before impact.

- The object ball moves off in the direction of a line joining the centres of the two balls at the moment of impact (which is observed in practice).

This is an extensive list of assumptions. However, unless we make them all, we cannot solve a two-dimensional collision for the velocities of the two balls after impact.

Example

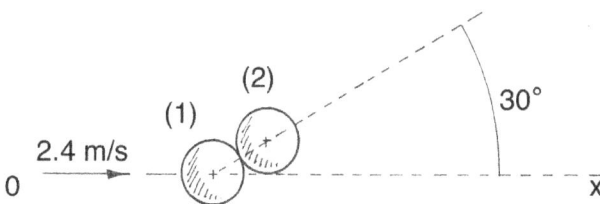

Billiard ball (1), moving at 2.4 m/s along a line o-x, strikes a stationary ball (2), of identical material and mass, with a glancing blow, such that the angle between the line joining their centres and line o-x is 30°.

Determine the velocities of the two balls after impact.

It is convenient to choose the path of the cue ball as the x-direction, so that this ball has a zero velocity component in the y-direction before impact.

Relying on all the assumptions listed above, we can state the following:

From equations (7) and (8) above:

In the x-direction we have: $\qquad 2.4 = v_{1x}' + v_{2x}'$ ....(i)

In the y-direction we have: $\qquad 0 = v_{1y}' + v_{2y}'$ ....(ii)

Since ball (2) moves away from the impact in the direction of the line shown:

$$v_{2y}' \div v_{2x}' = \tan 30° \quad ....(iii)$$

And, assuming no kinetic energy is lost in the collision: $\quad \frac{1}{2}mv_1^2 = \frac{1}{2}mv_1'^2 + \frac{1}{2}mv_2'^2$

In this equation, since the masses are assumed equal, ($\frac{1}{2}$m) cancels throughout.

Additionally, noting that $v_1'^2 = v_{1x}'^2 + v_{1y}'^2$ and similarly, $v_2'^2 = v_{2x}'^2 + v_{2y}'^2$,

the KE equation becomes: $\quad (2.4)^2 = v_{1x}'^2 + v_{1y}'^2 + v_{2x}'^2 + v_{2y}'^2$ ....(iv)

0.600 m/

1.039 m/s $\qquad$ 60°

$v_1' = 1.200$ m/s

1.039 m/s $\qquad$ $v_2' = 2.078$ m/s

30°

1.800 m/s

These four equations can be solved simultaneously for the four velocity components after impact. The magnitudes and directions of the velocities of the two balls respectively are shown here:

Observe that the final velocities of the two balls are at right angles to one another. This will be found to be the case for all such collisions, provided the masses of the two balls are equal, and the same set of assumptions is applied.

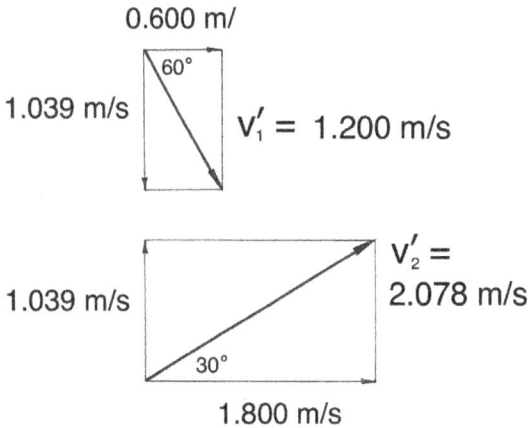

Exercise

A smooth 2 kg ball moving at 3 m/s on a horizontal surface strikes a glancing blow to a stationary 4 kg ball of the same elastic material, as shown below.

Determine the velocities of both balls immediately after impact. Assume the collision is perfectly elastic.

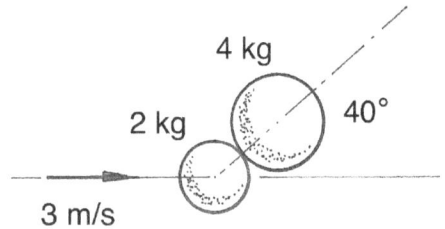

4 kg

2 kg $\qquad$ 40°

3 m/s

[The 2kg ball moves away from the impact at 2.075 m/s in direction E 71.67° S; the 4 kg ball moves away from the impact at 1.532 m/s, in a direction E 40° N]

Two dimensional vehicle collisions: working out what happened

The principle of the conservation of momentum can be used to analyse motor vehicle crashes. After a crash, it is often necessary to determine a close estimate of the speeds at which the vehicles were travelling before they collided. Police record tyre and skid marks to determine the paths covered by the vehicles involved. They gather evidence to find the point of impact, and how far the wrecks slid before coming to rest. If they have sufficient evidence, they can make an analysis similar to the following:

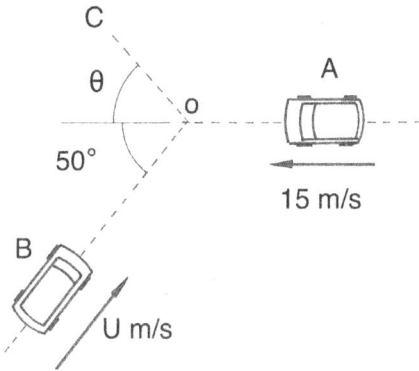

Example

Two cars collide on a horizontal road surface. Car A (1400 kg) is driving at 54 km/h when car B (1200 kg) comes out of a side road and hits it diagonally, head on, as shown. After the impact, the two cars, now entangled, move off together in the direction of o-C, which makes an angle θ with the original line of travel of car A.

Let the initial velocity of car B be U and the velocity with which the two cars move together after impact be V.

If the positive x-direction is chosen to coincide with the line of travel of car A before the collision, but to the right, then applying the equation of the conservation of momentum to the x- and y-directions respectively:

x-direction: $1400(-15) + 1200U \cos 50° = -2600 V \cos θ$ ....(1)

y-direction: $1200 U \sin 50° = 2600 V \sin θ$ ....(2)

This gives two equations with three unknowns. If we insert selected values for one of the unknowns, we can solve for the other two.

If we choose a set of values for U, we can solve for V and θ. Clearly, the greater the speed of car B, the greater will be angle θ, and the greater will be the velocity V.

Exercise: for the example above, confirm the values shown in the table below, and supply the missing figures.

Initial velocity of car B: U [m/s]	Velocity of tangled wreck: V [m/s]	θ [degrees]
10	6.21	34.68
15	6.42	55.63
20	7.39	73.14
25	8.86	85.73
30	10.64	
35		

The steady-state transfer of momentum by a fluid stream

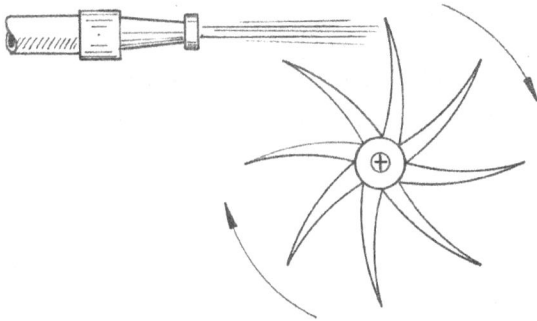

A stream of fluid, whether liquid or gas, can exert a force. We have mentioned examples like rockets, fire-hoses, and water that drives turbines or water-wheels.

Similarly, a stream of solid particles is also capable of exerting a force. Examples of this would include the stream of sand emerging from a sandblaster nozzle, and a stream of machine-gun bullets hitting a plate.

The amount of force a fluid stream can exert is determined as follows: We determine the force necessary to act *on* the fluid stream, in order to change its momentum at the rate that is specified. The fluid stream must then be exerting an equal and opposite force on its surroundings.

The force acting on a stream of fluid is equal to the rate of change of its momentum. This statement is the verbal equivalent of the equation we saw earlier in this chapter, namely:

$$dp/dt = F \dotfill (1)$$

If a stream of water emerging from a hose nozzle strikes a wall at a given velocity, and has its velocity reduced to zero by hitting the wall, then the stream has undergone a change in velocity.

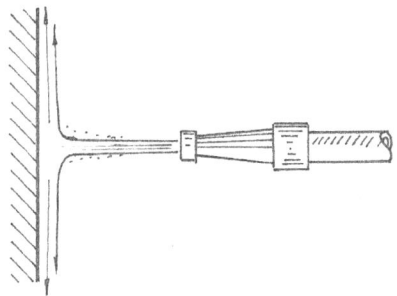

For this to happen, the wall has had to exert a force on the stream of water, therefore the water must have exerted the equal and opposite force on the wall.

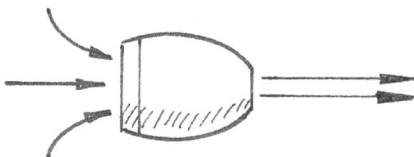

A jet engine takes in air at a certain speed, and forces it out at a greater speed. The engine has had to exert a force on the airstream to change its momentum.

If the jet engine exerts a force on the stream of air flowing through it, the airstream must be exerting an equal and opposite force on the engine, thus providing thrust.

If the fluid in question is flowing at a steady rate (without fluctuations), then the force required to accomplish the change in momentum:

$$F = dp/dt = (p_2 - p_1)/t = m(v_2 - v_1)/t = (m/t)(v_2 - v_1) \quad \text{.........................(9}$$

where m/t is the mass flow-rate, namely the mass of fluid passing a given point per unit of time. The mass flow-rate is usually designated by the symbol Q, and has units [kg/s].

Thus, the equation governing the external force needed to change the velocity of a stream of fluid or particles is:

$$\textbf{F} = \textbf{Q} (\textbf{v}_2 - \textbf{v}_1) \quad \text{..(10)}$$

Note: this is a vector equation. If all movement occurs in one straight line, the equation may be applied as if it were a linear algebraic equation. If movement occurs in two or three dimensions, the equation has to be applied separately in each of the principal directions.

Example

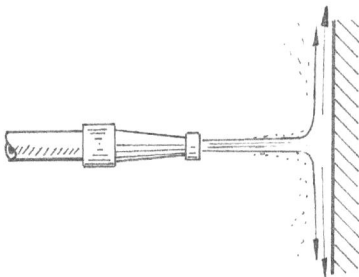

A hose discharges water flowing at 8 m/s to the right, with a flow-rate of 10 kg/s, against a wall, so that the forward velocity of the water is reduced to zero. What force does the stream of water exert on the wall?

Let the positive direction be to the right.

The force exerted *on* the water, to accomplish this change of momentum, is given by:
$$F = Q (v_2 - v_1)$$

$$= 10 (0 - 8) = -80 \text{ N, namely, 80 N to the left.}$$

Therefore the stream of water exerts a force of 80 N *to the right* on the wall.

Exercises on momentum transfer by a fluid stream

Question 1

A small rocket engine being bench-tested needs to have a thrust of 100 N. The exhaust gases emerge at a speed of 50 m/s relative to the rocket. At what mass flow-rate should the exhaust gases emerge from the rocket? [2 kg/s]

Question 2

A jet engine which is restrained from moving, fixed to a bench secured to the ground, draws in still air at a rate of 100 kg/s and discharges it with a velocity of 600 m/s. Determine the thrust developed by the engine. [60 kN]

Question 3

Water flows at a speed of 2.0 m/s down a concrete irrigation canal. 2 m³ of water passes any given point each second. This water passes through a grid placed across the canal to catch debris. In passing through the grid, the water speeds up to 2.2 m/s. (It has to move faster through the grid, as there is less cross-sectional area for it to occupy than in the open channel.) What force does the flow of water exert on the grid?   [400 N]

Determining the value of Q, the mass flow-rate of a fluid flowing in a defined stream

Consider a control volume of fluid flowing through a pipe at steady velocity v [m/s], taking t seconds to pass a given point in the pipe.

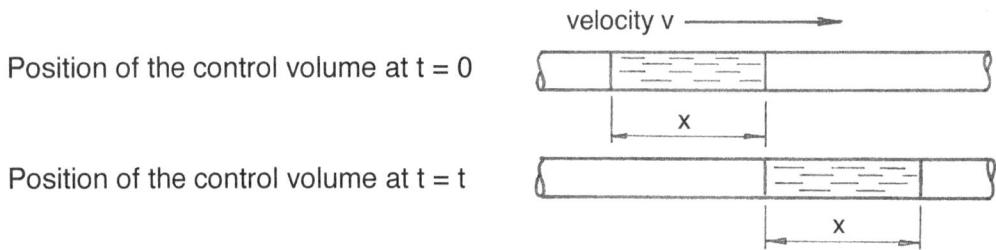

Position of the control volume at t = 0

Position of the control volume at t = t

If the length of this volume of fluid is x, then its velocity $v = x/t \therefore x = vt$

If the cross-sectional area of the flow is 'a', then the volume of this amount of fluid is equal to area × length = a(x) = a(vt)

Now, if the density of this fluid is ρ, then the mass of this volume of fluid is:

m = density × volume = ρ(avt)

So, the amount of mass passing the given point in time t is (ρavt)/t = ρav

$\therefore$ mass flow-rate **Q = ρav**...(11)

Example

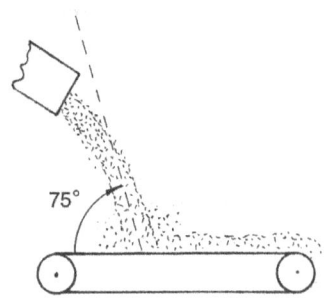

In an ore-processing plant, crushed gravel emerges from a chute at 90 litres per second, landing on a horizontal conveyor belt in a stream that makes an angle of 75° with the surface of the belt at the point of incidence, with an incident velocity of 5.40 m/s.

The belt speed is 2.0 m/s.

The density of this gravel stream (allowing for air spaces between the gravel chunks) may be assumed to be 65% of the density of the solid rock from which it is obtained, which is 2770 kg/m^3 on average.

Determine the force that the stream of gravel exerts on the belt at the point of incidence.

Solution: The approach needed is to determine the force needed to change the momentum of the gravel stream. This will be done using the equation $F = Q (v - u)$ applied separately in the x- and the y-directions.

The mass flow-rate $Q = 90$ [l/s] $\div 1000$ [l/m^3] $\times (0.65 \times 2770)$ [kg/m^3] $= 162.045$ kg/s

The components of the incident velocity are:

$v_{1x} = 5.4 \cos 75° = 1.398$ m/s and $v_{1y} = -5.4 \sin 75° = -5.216$ m/s

The final velocity of the gravel stream, once it is on the belt, is 2 m/s to the right. The y-component of this velocity is zero.

Components of force F exerted on the stream of gravel by the conveyor belt:

In the x-direction: $F_x = Q (v_2 - v_1)$

$= 162.045 (2 - 1.398) = 97.61$ N

In the y-direction: $F_y = Q (v_2 - v_1)$

$= 162.045 (0 - (-5.216)) = 845.2$ N

Recombine the two components of force F:

F = 850.8 N at an angle E 83.41° N (top left figure)

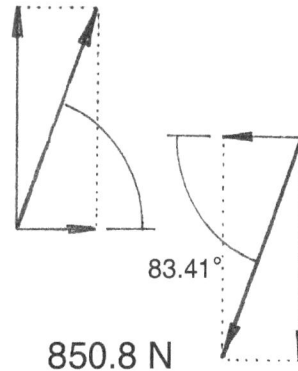

This is the force needed to change the momentum of the stream of gravel.

Hence, the force that the gravel stream exerts on the belt is equal and opposite to this, namely: 850.8 N at an angle W 83.41° S, as shown in the lower right figure.

Exercise

A hose discharges water at 50 litres/second from a nozzle with internal diameter 40 mm. The stream of water, moving horizontally, strikes a flat plate perpendicularly. Determine the magnitude of the restraining force required: a. to hold the plate still [1990 N] and b. to keep the plate moving steadily at 3 m/s to the right.   [1840 N]

Exercise

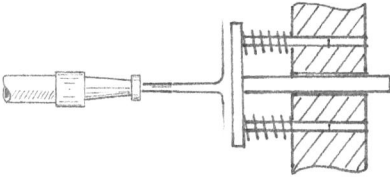

A stream of water is directed horizontally from a nozzle with an internal cross-sectional area of 400 mm².

This water hits a flat plate that is free to slide on guides, and which is cushioned by four coil springs, each of stiffness k = 1200 N/m.

Under steady flow, the springs compress enough for the plate to move 38 mm to the right, and remain in equilibrium in that position.

Determine the volumetric flow-rate of the water.   [8.542 litres/second]

Exercise

A ball of mass 1 kg is placed between four vertical rods such that it is constrained to move in a vertical path, and, due to a slight gap between the ball and the rods, the vertical movement of the ball is not restricted by any friction with the rods.

A jet of water is directed upward at the underside of the ball, and is sufficiently strong to keep the ball in equilibrium at a height of 1.2 m above the nozzle, whose internal diameter is 20 mm.

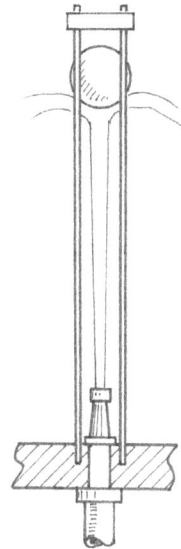

Assume the vertical component of the velocity of the stream of water is reduced to zero immediately after impact with the ball.

Determine the mass flow-rate of the water emerging from the nozzle.   [2.111 kg/s]

Example

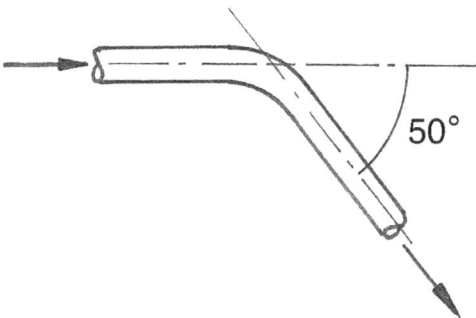

Illustrated here is a plan view of a water pipe with a bend in it, lying in a horizontal plane.

Water flows through the pipe with a mass flow-rate of 10 kg/s, at a constant speed of 5 m/s.

Determine the magnitude and direction of the force that the water exerts on the pipe at the bend.

The force that must be exerted *on the water-stream* to change its momentum is given by:

$F = Q (v_2 - v_1)$ in each of the principal directions.

Let the original direction of flow before the bend coincide with the positive x-axis.

In the x-direction: $F_x = 10(5 \cos 50° - 5)$

$= -17.86$ N

In the y-direction: $F_y = 10(-5 \sin 50° - 0)$

$= -38.30$ N

The components of the force that the water exerts on the pipe are directly opposite to these, as shown here. Also, note that the direction of F bisects the angle of the bend. You might set up your own similar exercise with a different angle in the pipe, and see if this pattern still holds.

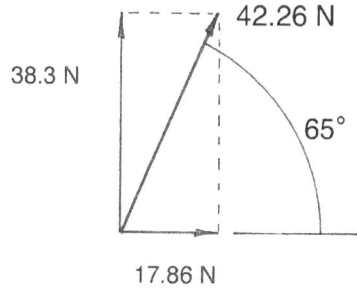

Exercise

A stream of sand particles emerging horizontally from a sandblaster nozzle at a speed of 10 m/s, with a mass flow-rate of 4 kg/s, impinges on a steel plate suspended from hinges that experience negligible friction.

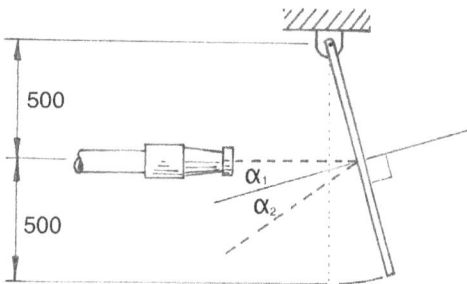

The plate originally hangs vertically.

The centre-line of the stream of sand particles coincides with the original position of the centre of gravity of the hanging plate, 500 mm below the hinge line.

Once the stream of sand particles is flowing steadily, the plate makes an angle $\theta = 15°$ with the vertical.

Ignore the effect of gravity on the stream of sand particles. Assume that the sand particles bounce off the plate such that the angle of incidence, α_1 is equal to the angle of departure, α_2. Determine the mass of the plate.   [31.51 kg]

Example

A human-powered water craft uses for propulsion a pedal-driven pump which takes in water at the front of the craft and pushes it out of a rear-facing pipe at 10 m/s relative to the craft.

Assume the resistance to motion of the craft is constant at 60 N, irrespective of speed. If

189

it is desired to make the craft move forward at a speed of 5 m/s relative to the water surface, determine:

- the number of litres per second that must emerge from the pipe, and
- the power that must be generated by the person pedalling.

Solution: This exercise is about determining the force needed to change the momentum of a stream of water, namely the water that is taken in at the front of the craft and pushed out behind it. The propulsion method is similar to that of a jet engine, employing water instead of air.

This craft is moving at constant velocity, and is therefore in equilibrium. The thrust obtained from pushing backwards a stream of water will be equal and opposite to the force needed to change the momentum of that stream.

The thrust must also equal the sum of the resistances to motion of the craft, given as 60 N.

Consider the forward direction of motion of the craft to be positive, and define the initial and final velocities of the stream relative to the craft. Hence, $u = -5$ m/s and $v = -10$ m/s.

If the mass flow-rate of the stream is Q [kg/s], then the force needed to change the momentum of the stream:

$$F = Q(v - u) \quad \therefore -60 = Q(-10 - (-5)) \quad \therefore Q = 12 \text{ kg/s, hence 12 litres per second}$$

The power output of the person pedalling, ignoring friction loss in the pedalling mechanism and the pump: $P = Fv =$ (thrust force) × (velocity of craft relative to the main body of water) = 60 N × 5 m/s = 300 W

Notes and questions connected with this example

1. For how long do you think a normally athletic person could keep up this rate of energy expenditure?

2. Would the task be easier or more difficult in sea water, compared with fresh water?

3. Why did we not need to specify the mass of the craft or rider?

4. The arbitrary figure of a constant 60 N that was given for the resistance to motion of the craft was provided to simplify the solution to this problem. More realistically, the resistance to motion depends on which other factors?

5. What design features would you employ to reduce the resistance to motion of such a craft?

Summary: the main points of principle in this chapter

The Law of Conservation of Momentum states that the total momentum of a system of objects is 'conserved'.

This means that, in the absence of any external forces acting on the system of objects, the total momentum of that system remains constant.

This law allows us to analyse the forces and velocities that occur in connection with collisions and with the steady-state behaviour of a stream of fluid or solid particles.

Recall that the principle of the conservation of energy applies to smooth transfers of energy, *in the absence of collisions*. The conservation of energy cannot be used to analyse collisions, because too much energy is 'lost' in any impact.

Therefore, to analyse any impact situation, the principle of conservation of momentum must be applied.

Chapter 17

Relative velocity

Definitions of absolute, relative and resultant velocity
Ways of describing the directions of velocity vectors
Determining a relative velocity: the trick of moving the plane
Methods of solving velocity diagrams
The relative path
Intercept situations
Relative velocity applied to simple mechanisms

Definitions of Absolute, Relative, and Resultant Velocity

Absolute velocity, also called actual or true velocity

Picture the most tranquil circumstance you can think of. Suppose you are lying on a beach, under a tree, on a perfect summer's morning. It is one of those days when nothing seems to move. Even the wavelets are barely lapping. The air is still, and you are half-asleep, with your eyes closed. You feel as if you are completely motionless. But, are you?

Well, consider this: you are on the surface of the Earth, which is spinning on its axis such that your peripheral speed, depending on your latitude, could be anything up to approximately 1668 km/h at the equator.

The Earth is also moving in an orbit around the sun, at a speed of 107 000 km/h. Besides which, the whole solar system, with you in it, is scorching through space with an astonishing combination of velocities. According to the Stanford Solar Center, our sun is moving toward another star in our galaxy at 72 000 km/h. To add to the picture, our entire solar system is moving in an orbit caused by the spin of our galaxy, such that the solar system as a whole has a peripheral speed on this orbit of 792 000 km/h. As if that were not enough, the entire galaxy is moving through space, away from the Big Bang location, at 2.1 million km/h. So much for your sense of tranquillity!

Lying quietly on that beach, you are not moving relative to the Earth. But, you definitely *are* moving, relative to Saturn, the moon and any star you care to name. Likewise, *they* are all moving, relative to *you*. We are compelled to conclude that there is no such thing as an 'absolute' velocity. The velocity of any object can only be defined relative to another object.

Since our home is the Earth, and we move around on it, and most of our engineering is conducted on or close to its surface, we often refer to a velocity relative to the Earth as an 'absolute', 'actual' or 'true' velocity. You can see why this choice of terms needs to be used with caution. We need to remain mindful that 'true velocity' is *chosen* to mean 'velocity relative to the Earth'.

Relative velocity

This is the *apparent* velocity that one object has, as seen from another. When you are in a car doing 60 km/h, and another car is coming down the road towards you, travelling at 80 km/h, the two vehicles are approaching each other at 140 km/h.

80 km/h   60 km/h

Neither car is moving with a speed of 140 km/h relative to the road. However, to the occupants of both cars, the other car appears to be approaching at 140 km/h. We say the other car's speed, relative to our car, is 140 km/h.

If we take into account their directions of travel, to speak of velocities rather than speeds, we see that the velocity of A, relative to B, is exactly equal and opposite to the velocity of B, relative to A. We will see later that this statement remains true, even when the two vehicles are not moving in the same straight line.

Relative velocity is very important in mechanical engineering. A knowledge of how to determine the magnitude and direction of a relative velocity is of use in many situations, for example:

- the velocity of one machine part relative to another, when both are moving,
- the velocity of a machine tool tip relative to the material being processed,
- the velocity of a fluid relative to a turbine blade or propeller that is moving within the fluid, and
- The velocity of a stream of solid particles dropping onto a conveyor belt, relative to the belt.

In this chapter, we will get to understand relative velocity through examples involving the motion of cars, aircraft and ships relative to one another, in the same plane. We will also apply this understanding to some elementary mechanical situations.

However, before being able to determine the value and direction of a relative velocity, we need to understand the concept of a *resultant* velocity.

Resultant Velocity

If an object is moving in a medium which is also moving relative to the Earth, that object's velocity relative to the earth is the *resultant* of the two velocities.

For example, suppose you are in a rowing boat, trying to cross a river, starting a point A, aiming at a point B on the opposite bank. You might be rowing at 3 km/h, at right angles to the current, but the current itself may be moving at 5 km/h. You will not land on the opposite bank at point B. You will land at a point C, downstream from there.

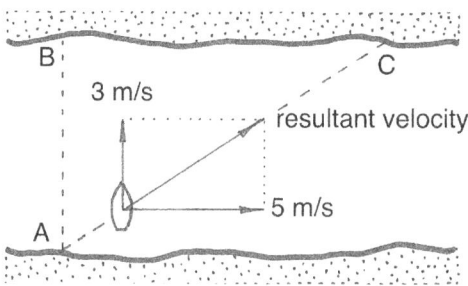

This is because your 'actual' velocity (namely, your velocity relative to the Earth) is the vector sum (the resultant) of the two velocities mentioned, and the direction in which you travel is the direction of that resultant velocity.

Further examples of resultant velocity

1. You are walking up an escalator at 1 m/s. The escalator is moving upwards at 0.5 m/s. Your velocity relative to the building (Earth) is the combination (or resultant) of these two velocities, namely 1.5 m/s.

2. A spy trying to escape runs forwards along the top of a moving train carriage, the better to dodge the bullets being fired by guards on the ground. If he runs at 6 m/s and the train is moving at 18 m/s, his resultant velocity relative to the ground is 24 m/s. If he were to run towards the back of the train, his resultant velocity would be only 12 m/s forwards, making him an easier target.

3. An aircraft is attempting to fly directly from A to B while a cross-wind is blowing.

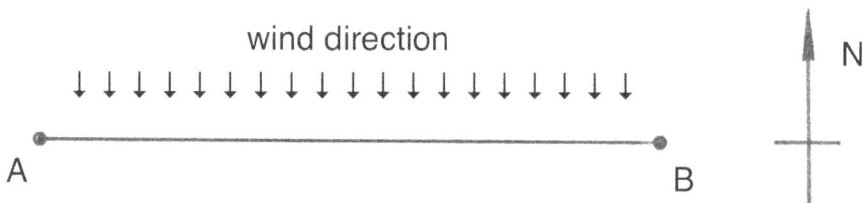

The pilot has to compensate for the tendency of the wind to blow the plane to the south, by steering a course slightly to the north of the direction in which he needs to fly.

195

He needs the ground-speed of the aircraft to be directed from A to B. This means that the actual path travelled will be a direct line from A to B, and an observer on the ground anywhere along line AB will see the aircraft passing directly overhead.

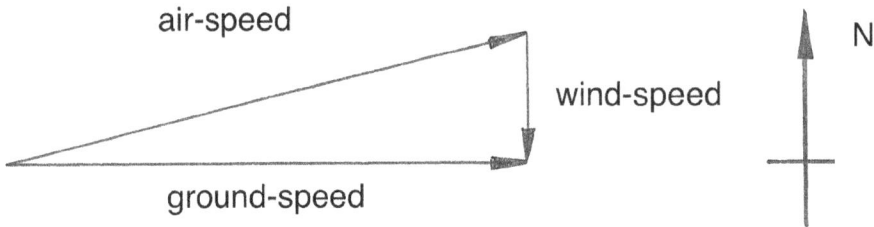

The ground-speed is the resultant of the two velocities, namely the velocity of the aircraft in the air and the velocity of the air (the moving medium) itself.

The time taken to do the journey is given by
(distance AB on the ground) ÷ (ground-speed).

Ways of describing the directions of velocity vectors

In order to draw and analyse velocity diagrams pertaining to vehicles such as cars, ships and aircraft moving in the 'plane' of the earth's surface, we are going to need ways to specify the *directions* of velocity vectors.

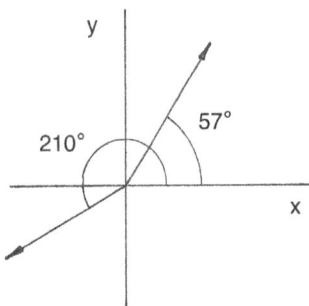

In common mathematical practice, when using 2-D rectangular co-ordinates, the direction of a vector is usually specified in terms of the angle measured anticlockwise between the x-axis and the given vector.

While this method is always valid and useful, it is not customarily used to describe the directions of velocity vectors that occur in the 'plane' of the Earth's surface.

There are three different ways in which directions are customarily specified for such vectors.

1. Specifying an angle in a quadrant between two principal directions

Velocity vectors can be drawn showing their direction relative to the four principal directions, north, south, east and west.

To describe the direction of a velocity vector, we can look at the quadrant in which that vector lies, and state the angle that occurs between that vector and the two adjacent principal directions between which it lies.

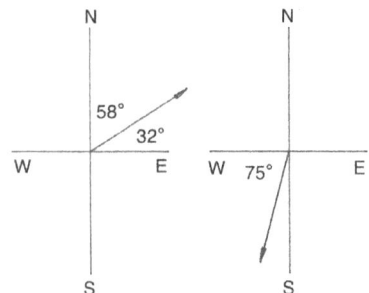

For example, in the first illustration here, if you

start at east and rotate through 32° towards north, you will arrive at the direction in which this vector is pointing. This direction is described as E 32° N.

The same direction can be described by choosing to start at north and rotating through 58° towards east, in which case it is designated as N 58° E. These two expressions mean exactly the same thing. The point is, this direction is fully defined, as it is referenced to two of the principal directions.

The direction of the second velocity vector illustrated above could be described as either: S 15° W or W 75° S.

2. Specifying a compass bearing

To describe the direction of a velocity vector originating at the observer, or to specify the direction from the observer to an observed point, it is common practice in navigation to describe the number of degrees rotated through, starting at North and moving clockwise.

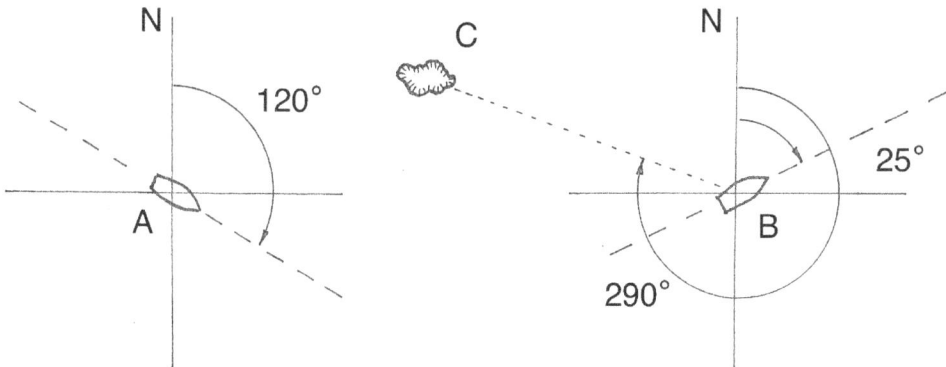

In the illustrations here, ship A is proceeding on a course with a bearing of 120°. Ship B is proceeding on a bearing of 65°. The rock C, as seen from ship B, will be found on a bearing of 290°.

3. Specifying a relative bearing

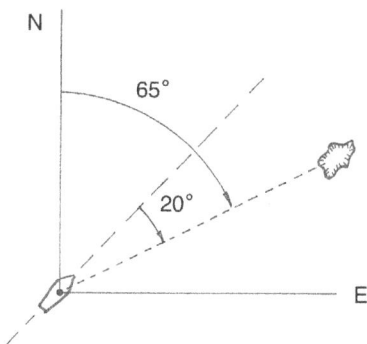

For a relative bearing, the reference direction is not north, but is the forward projection of the centre-line of one's own vessel. So, you might be on a vessel moving in any given direction, such as north-easterly (namely, on a compass bearing of 45°) when you spot an island that can be seen by looking 20° to the right of the ship's line of symmetry. The relative bearing of the island, seen from your vessel, would be 20°.

The compass bearing of the line of sight to the island would be 45° + 20° = 65°.

A fourth method, similar to specifying a relative bearing, but less precise, is the convention of indicating directions by comparing them to the hour divisions on a horizontal clock face, where 12 o'clock represents 'straight ahead' and 6 o'clock represents 'directly behind us'. This method is used in situations where time is short and colleagues need to respond quickly, such as in aerial battles.

Exercise in determining both compass bearings and relative bearings

At a given instant, three ships are in the positions shown. Determine and write down the true bearing, and the relative bearing, of each ship as seen from each of the other two ships.

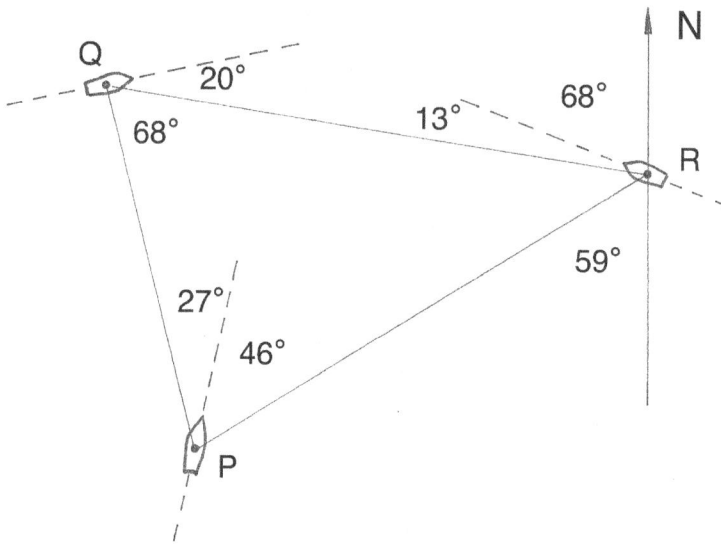

Exercises: Determining Resultant Velocities

Question 1

A boy is standing on the back of a truck moving at 72 km/h along a straight stretch of road. Using a catapult, he shoots a stone horizontally from that position. The stone moves away from the catapult at 20 m/s. What is the initial velocity of the stone relative to the road, if he shoots it:

  a. Directly ahead? [40 m/s]
  b. Directly backwards? [0 m/s]
  c. 20° to the right of the direction the truck is moving? [39.39 m/s at 10° to the road, namely on a relative bearing of 10°.]
  d. At right angles to the road, to the left of the truck? [28.28 m/s at 45° to the road, namely on a relative bearing of 315°.]

Question 2

A hot air balloon rises in air with a vertical speed of 2 m/s. The wind is blowing from the west, with a speed of 1.2 m/s. If the balloon is released from point P on the ground at 08:00,

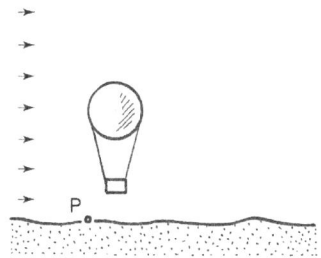

a. What is the resultant velocity of the balloon? [2.332 m/s, 59.04° from the horizontal, moving east.]

b. How far from point P will the balloon be at 08:10? [1399 m]

Question 3

A man who can row his boat at 6 km/h in still water, wishes to cross a river which is 660 m wide, and flowing at 4 km/h. If he rows at right angles to the stream:

a. What is his resultant velocity? [7.211 m/s at 56.31° to the bank]

b. How far downstream will he land? [440 m]

c. How long will it take him to cross the river? [1 min 50 sec]

If, instead, he wishes to cross by the shortest route:

d. In what direction should he head? [upstream at 41.81° to the line perpendicular to the banks]

e. What will be the magnitude of his actual velocity? [4.472 m/s]

f. How long will it take him to cross? [2 min 27.6 sec]

Question 4

An aircraft with a cruising speed of 180 km/h is to fly from base A to base B, which is 250 km to the east. There is a wind of 50 km/h blowing from the north.

a. In what direction should the pilot head so that the resultant velocity of the aircraft is directed from A to B? [on a bearing of 73.87°]

b. What is the magnitude of this resultant velocity? [172.9 km/h]

c. How long will it take the aircraft to get to B? [1:26:45]

Determining a relative velocity: The trick of moving the plane

Suppose two vehicles, A and B, are moving, each in its own straight line, on different paths, on a flat plane.

We want to see what the velocity of B appears to be, as seen from the point of view of an observer in vehicle A. This is called the velocity of B relative to A, designated V_{BA}.

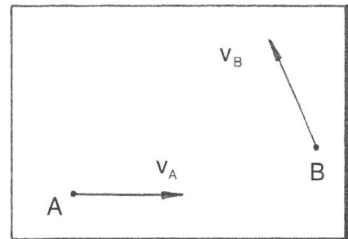

For you, the external observer, to see the velocity of vehicle B, as it would be seen from vehicle A, imagine that you can impart a movement to the plane. If you move the plane in such a way to make vehicle A appear stationary in front of your eyes, then you have the same viewpoint as vehicle A.

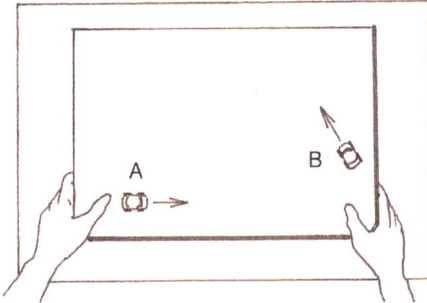

This can be demonstrated quite easily, using a tray or a light board on which two small wind-up toy cars are placed. One person holds the sides of the tray which is resting on, and free to slide on a horizontal table surface.

Two others wind up the springs of the toy cars, and at a signal from the tray-holder, they place their cars on the tray, and allow the cars to run, each in its respective path.

The person holding the tray then slides the tray in such a way that car A is made to stand still relative to the observers. All three observers then note the apparent direction of the path of car B.

In order to see what the velocity of B appears like, as seen from A, you have had to make vehicle A stand still in front of your eyes. To do this, you needed to move the plane with a velocity equal and opposite to that of vehicle A. In other words, you gave *the whole plane* a velocity $(-V_A)$. *Note: the vector $(-V_A)$ does not have a negative value: it is simply the opposite vector to V_A.*

So, everything else on the plane has now acquired a velocity $(-V_A)$.

This means that vehicle B now has a resultant velocity, equal to the vector sum of its velocity on the plane, and the velocity that you imparted to the plane, namely the vector sum of (V_B) and $(-V_A)$.

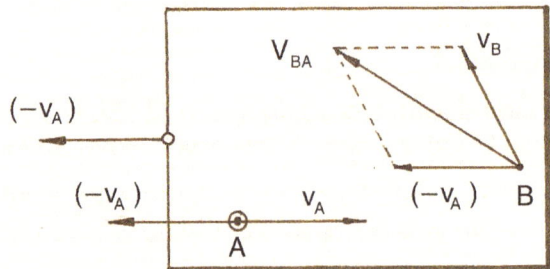

This resultant velocity is how the velocity of vehicle B appears from vehicle A. It is therefore V_{BA}.

So, the velocity of B relative to A is the resultant of V_B and the *opposite* vector to V_A.

This is expressed by the vector equation: $V_{BA} = V_B + (-V_A)$

Note: this is *not* an *arithmetic* equation. It can only be interpreted *arithmetically* in the extremely elementary case where V_B and V_A occur in the same straight line. This vector equation may be applied *all* instances where a relative velocity needs to be determined.

The order in which the subscripts appear in this equation is important, and is *always* the same: if the LHS has subscript BA, the RHS will have subscripts: B followed by A.

For example, if you need to determine the relative velocity of an object C, as seen from an object D, then the form of the equation to use is: $V_{CD} = V_C + (-V_D)$

The vector equation represents a velocity vector diagram, on which the resultant vector V_{BA} is the diagonal of a parallelogram whose adjacent sides are V_B and $(-V_A)$.

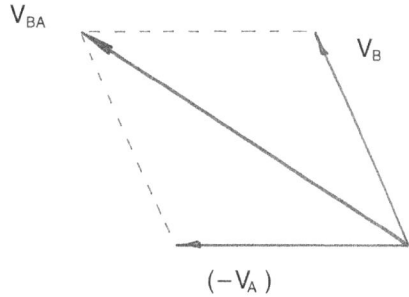

V_{BA} V_B $(-V_A)$

Methods of solving velocity diagrams

Three ways:

- graphically, by constructing the parallelogram to scale, and reading the value and direction of the resultant off the diagram,
- using a roughly drawn diagram, applying geometry and trigonometry to determine the required magnitudes and directions, and
- by the summation of vector components in the x- and y-directions.

Example

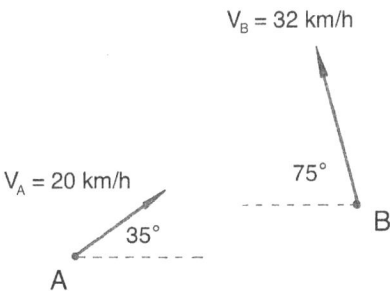

$V_B = 32$ km/h

$V_A = 20$ km/h

75°

35°

A

B

At a given moment, two ships near one another are proceeding on straight courses, with the respective velocities shown. Determine the velocity of ship A relative to ship B.

Solution:

1. Write down the vector equation: V_{AB} is needed, thus, taking into account the order of the subscripts, we write: $V_{AB} = V_A + (-V_B)$.
2. Construct a parallelogram of velocity vectors that corresponds with this equation. The two adjacent sides are respectively the vectors V_A and $(-V_B)$, and the diagonal represents the resultant of these, namely V_{AB}.
3. Determine V_{AB} by any one of the following three methods, as demonstrated below:

a. **Graphically, to scale:** draw the diagram as accurately as possible, and measure from the diagram the magnitude and direction of the resultant.

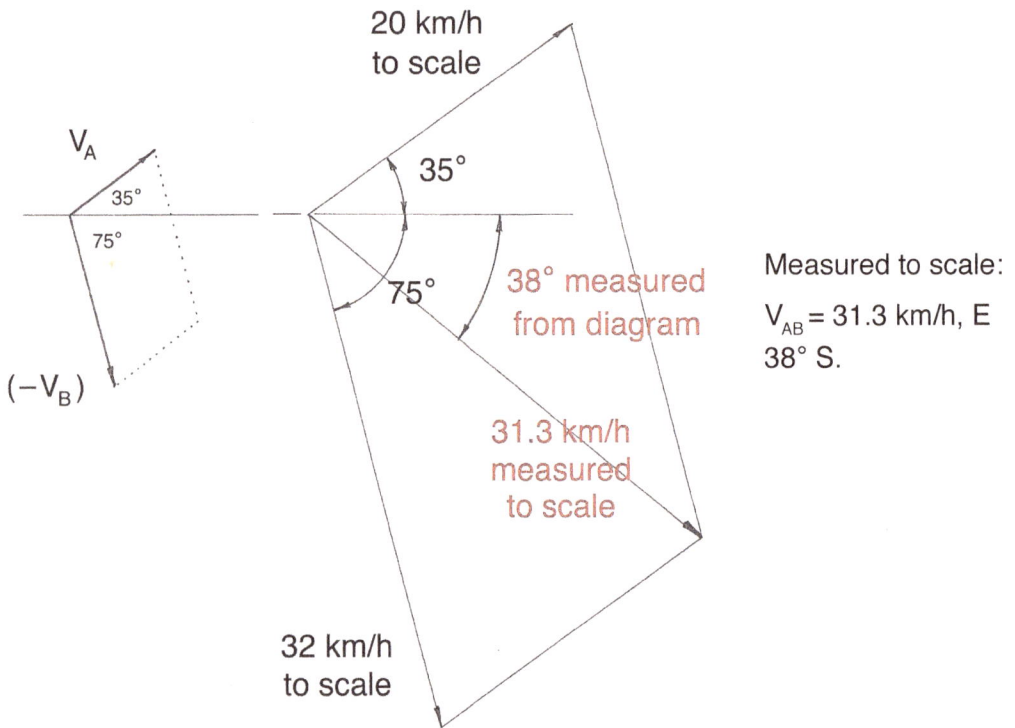

20 km/h
to scale

V_A

35°

35°

75°

75°

$(-V_B)$

38° measured
from diagram

31.3 km/h
measured
to scale

32 km/h
to scale

Measured to scale:

V_{AB} = 31.3 km/h, E
38° S.

b. Solving the velocity diagram using geometry and trigonometry:

Angle $\alpha = 180° - 110° = 70°$

Use the cosine rule to solve the lower triangle:

$|V_{AB}|^2 = 32^2 + 20^2 - 2(32)20) \cos 70°$

Therefore $V_{AB} = 31.40$ km/h

In the lower Δ : by the sine rule:

$\sin \beta /20 = \sin 70°/31.40,$

hence $\beta = 36.76°$

From which we deduce that the direction of V_{AB} must be E 38.24° S

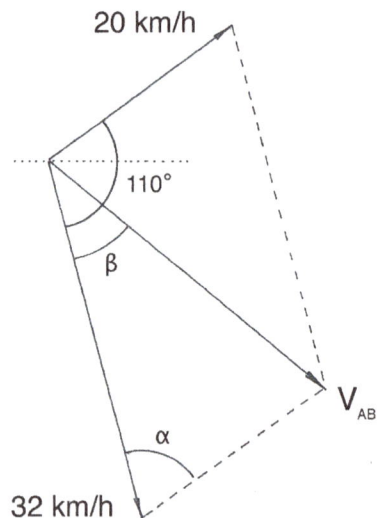

20 km/h

110°

β

V_{AB}

α

32 km/h

c. Determining the resultant by summing rectangular components:

Vector	x -component	y-component
V_A	20 cos 35°	20 sin 35°
$-V_B$	32 cos 75°	−32 sin 75°
Summation	24.6653	-19.4381

Recombine these two components:

$V_{AB} = \sqrt{(24.6653^2 + 19.4381^2)} = 31.40$ km/h

and angle $\theta = \tan^{-1}(19.4381 \div 24.6653) = 38.24°$

which exactly replicates the previous result.

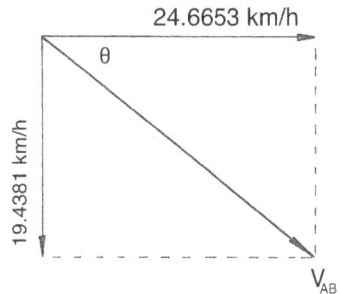

Any of the above three methods may be used. The graphical method, although the quickest, is subject to potential inaccuracies, which may or may not be significant. Careful drawing of velocity diagrams to scale usually yields answers accurate to within 1% of exact values. For most purposes this is accurate enough.

If a more accurate answer is desired, it is recommended that method (b) or (c) should be used, with method (a) applied as a quick check.

Exercises: Basic relative velocity diagrams

Question 1

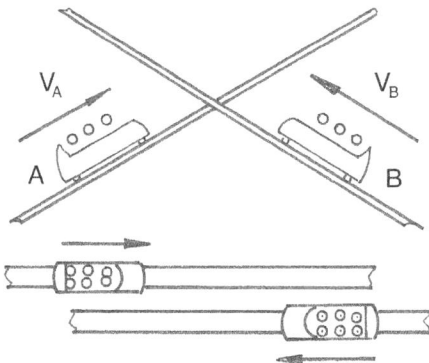

On two adjacent sections of roller-coaster track (both inclined at 30° to the horizontal) two cars pass each other, as shown. Car A is moving at 20 m/s and car B is moving at 30 m/s.

Determine the apparent velocity of B as seen from A.

[156.9 km/h, at 6.59° above the horizontal]

Question 2

The diagram shows a plan view of some linked parts in a mechanism, all constrained to move in the same horizontal plane. Part G is moving due north at 20 m/s. At the same instant, part H is moving due west at 35 m/s.

Determine the velocity of H with respect to G. Illustrate this clearly on a vector diagram.   [40.31 m/s, W 29.74° S]

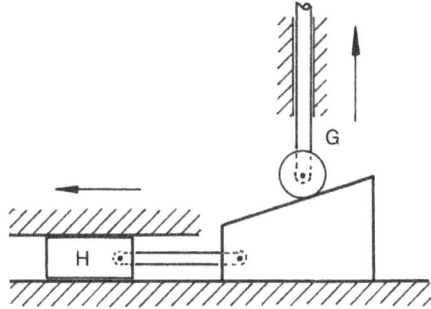

Question 3

Two cars are travelling on a flat plane, each with uniform velocity. Car P is moving due south at 40 km/h. To an observer in car P, car Q appears to be moving north-east at 30 km/h.

What is the actual velocity of car Q?   [28.34 km/h, S 48.47° E]

Question 4

An aircraft needs to travel from point A to point B, which is 225km due east of point A. The aircraft flies at a cruising speed of 162 km/h.

a. If there is no wind, how long will it take? Ignore the reduction of speed during take-off and landing. [1:23:20]
b. If there is a wind blowing at 54 km/h from the north-east, in what direction should the pilot head so that he travels in a straight line towards his destination? [E 13.63° N]
c. In this case, what is his velocity relative to the Earth (ground-speed)? [119.3 km/h]
d. How long will it take him to get there?   [1:53:12]
e. If, instead, the wind was blowing from the south-west at 36 km/h, answer the same questions as in b, c and d above.   [E 9.041° S; 185.4 km/h; 1:12:48]

The Relative Path

When two objects are moving with different velocities in the same plane, to each object, the other will appear to move along a path that is not a real path on the plane, but is an apparent path, relative to the moving observer. This is called the relative path.

The path that an object appears to be moving on lies in the same direction as the velocity it appears to have. So, the relative path and the relative velocity have the same direction.

The relative path can be used to determine the closest distance that will separate the two moving objects if they continue on their present courses. It can also be used to determine the conditions required for one of the moving objects to intercept the other.

Example

Two ships at sea are in the positions shown, with the velocities shown, at 12 noon.

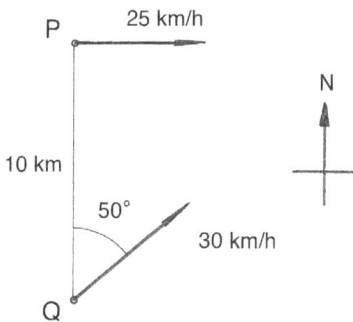

a. Determine the velocity of P relative to Q.

b. Show on a diagram the relative path that P appears to follow, as seen from Q.

c. What will be the closest distance between them?

d. At what time will they reach the point of closest approach?

Solution: a. $V_{PQ} = V_P + (-V_Q)$ hence the vector diagram below:

By the cosine rule:

$V_{PQ}^2 = 25^2 + 30^2 - 2(25)(30) \cos 40°$

$\therefore V_{PQ} = 19.389$ km/h

By the sine rule:

$(\sin \alpha)/30 = \sin 40°/19.389$

$\therefore \alpha = 84.024°$

$\therefore$ Direction of V_{PQ} is E 84.02° S

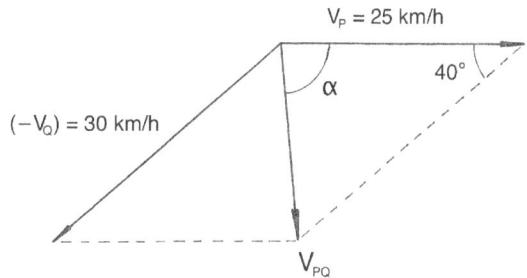

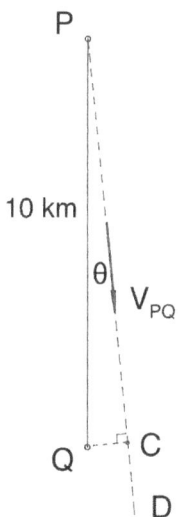

b. The relative path of ship P, as seen from ship Q, lies in the same direction as the relative velocity V_{PQ}. This path is indicated by the line PD on the diagram. In determining the velocity of P relative to Q, we have effectively stopped ship Q, as was done in the 'trick' of moving the plane, with the two toy cars. Seen from ship Q, ship P appears to start at its original position, and proceed along the relative path PD.

c. The shortest distance between the two ships will be QC, such that QC is perpendicular to PD.

Angle $\theta = 90° - 84.024° = 5.976°$

The shortest distance, QC = 10 km sin 5.976° =1.041 km

d. The time to reach the point of closest approach is the time taken for ship P to travel the relative path PC at the relative velocity.

Distance PC = 10 cos 5.976° = 9.946 km

If the speed remains constant, s = ut

$\therefore$ t = s $\div$ u = 9.946 km $\div$ 19.389 km/h

= 0.51297 hours = 30 min 47 seconds $\quad\therefore$ the ships reach this position at 12:30:47

Example

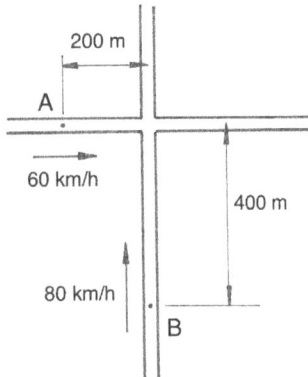

At a given instant, two cars in the positions shown are moving towards a crossroads with the given velocities.

If the cars do not change their speeds, determine the time it takes to get from the initial position (shown here) to the point at which they are the closest to one another.

Indicate on a diagram the actual positions of the cars on the road, at the instant when they are closest to each other.

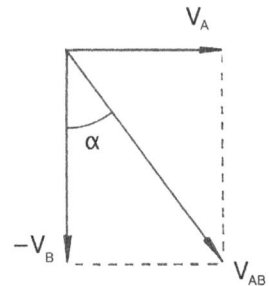

First, determine V_{AB} from the general equation for relative velocities: $V_{AB} = V_A + (-V_B)$ and draw the vector diagram. From this diagram, we deduce that V_{AB} = 100 km/h and angle α = tan^{-1}(60/80) = 36.87°

Use as reference points the positions that the two cars occupy initially. From the given dimensions, we can determine that the distance between the two cars in the initial position is 447.21 m, and that the line joining them makes an angle of 26.57° with a north-south line.

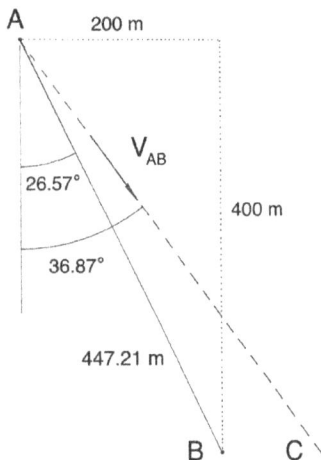

Now draw the actual initial positions of the two cars, to scale, as well as the direction of the relative velocity, V_{AB}.

From car B's point of view, it appears as if car A starts at point A, and moves along line AC with the relative velocity V_{AB} (namely 100 km/h).

It is important to remember that the relative path is *not* an actual path on the plane on which the two cars are moving.

Car A does not actually move along line AC. It only *appears* to do so, as seen from car B, which is also moving. Hence, the relative path is also moving, and cannot be traced on the ground.

Below is the relevant part of the above diagram that is needed to determine the shortest distance between the two vehicles.

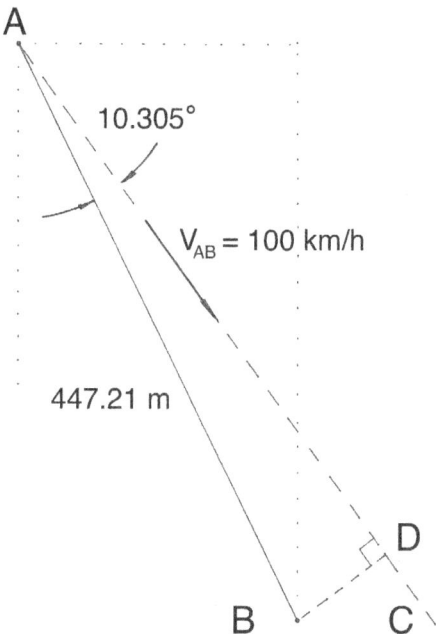

Car A will be closest to car B when it gets to the point D.

The shortest distance between the two cars will be BD, where

BD/447.214 m = sin (36.87° – 26.57°)

∴ BD = 447.214 × sin 10.305°

∴ BD = 80.00 m

How long does it take to get to the position of closest approach? This will be the time it takes for car A, travelling at the relative velocity of 100 km/h, to cover distance AD on the relative path.

Now, AD/447.21 = cos 10.3°

∴ AD = 440.0 m

Time taken = s ÷ u

∴ t = 0.440 km ÷ 100 km/h = 0.0044 hours = 15.84 seconds

In the diagram: A, 10.305°, V_{AB} = 100 km/h, 447.21 m, D, B, C

Where are the two cars actually, on the road, at the moment when they are closest to one another? Each one has been moving at its steady velocity along its real path, for the time of 15.84 seconds.

So, car A has covered an actual distance:

s = (60 ÷ 3.6) m/s x 15.84 sec = 264 m, and is therefore 64 m to the east of the crossroads.

Likewise, car B has covered:

s = (80 ÷ 3.6) m/s x 15.84 sec = 352 m, and is therefore 48 m south of the crossroads. Confirm using the theorem of Pythagoras that the distance between them must be $\sqrt{(64^2 + 48^2)}$, namely 80 m.

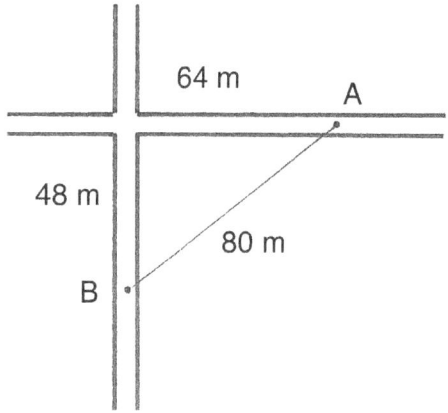

In the diagram: 64 m, A, 48 m, 80 m, B

This corresponds with the value we obtained using the relative path method.

Exercises on Relative Paths

Question 1 *(This exercise allows us to verify the result given by the method of relative paths by plotting the actual positions of moving objects at given times.)*

On the diagram below are shown three airfields, A, B and C. Airfield B is 60 km due east of A, and C is 50 km due north of B. Aircraft P starts at airfield A and heads towards B with a cruising speed of 180 km/h. At the same time, aircraft Q starts at B and heads for C, doing 120 km/h. Ignore the short period of initial acceleration during take-off. Assume velocities remain constant throughout.

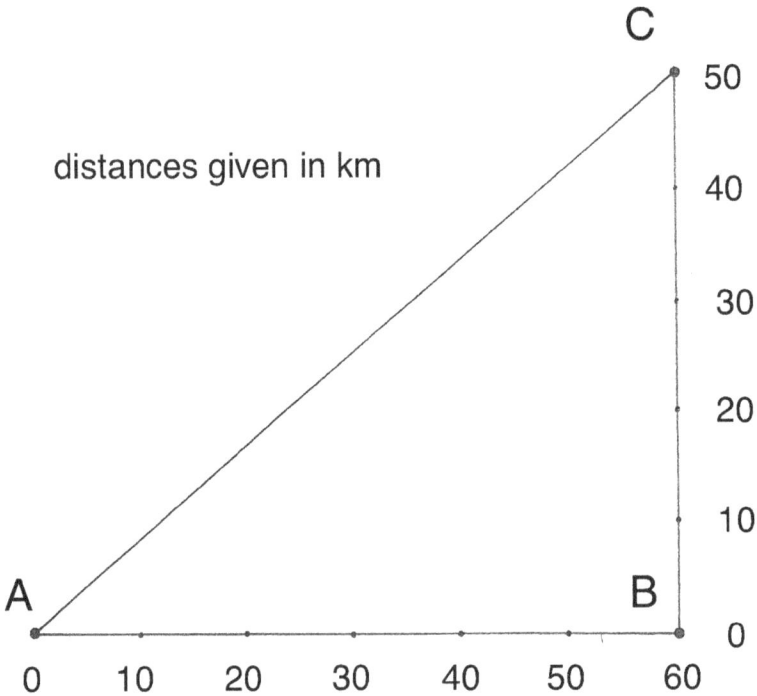

distances given in km

Plot their actual positions on this space diagram, at five-minute intervals. Join their actual positions at each respective time-interval by a dotted line.

Estimate the closest distance between them, and note where they are in relation to one another when they are at the point of closest approach.

Now determine, separately, the relative path of aircraft Q as seen from P, and, from the relative path diagram, determine the closest distance between the two aircraft. Again, note where they are in relation to one another when they are at the point of closest approach, and how long after starting they reach this position. Compare these results with those obtained by plotting the lines on the original diagram.

[Closest distance between them is 33.28 km; 13 min 50.8 sec after starting]

208

Question 2

At time 10:00:00 two ships are in the positions shown, with the velocities shown. B is exactly 10 km east of A.

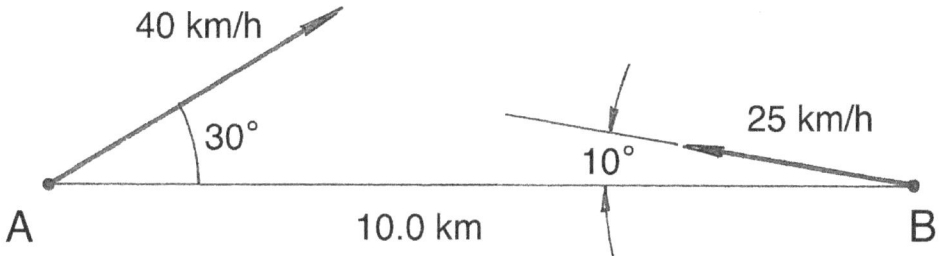

a. Using a graphical method, determine the velocity of A relative to B. (Draw your vector diagram quite large, and to scale) [V_{AB} = 61.30 km/h]

b. Draw to scale the relative path of A, as seen from B. [The path runs E 14.8 ° N]

c. What will be the closest distance between them?   [2.55 km]

d. At what time will they reach the point of closest approach?   [10:09:28]

Question 3

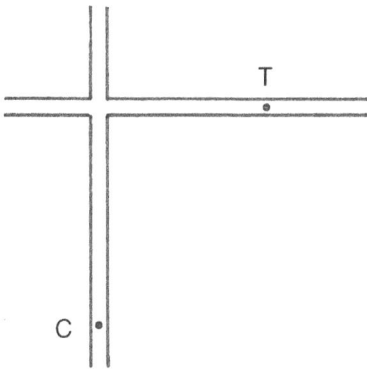

A car 'C' and a truck 'T' are approaching a rectangular intersection of two straight roads, respectively running north-south and east-west.

At a given instant the car is 200 m from the intersection, going north at 108 km/h. At the same instant, the truck is approaching from the east, 160 m from the intersection, at 72 km/h.

a. Determine the velocity of the truck relative to the car. [129.8 km/h or 36.056 m/s; E 56.31° S]

b. Draw to scale the relative path of the truck, as seen from the car.

c. If they do not change their speeds, what will be the closest distance between them?   [22.19 m]

d. At what time will they reach the point of closest approach?   [after 7.077 sec]

Group Tutorial Questions on Relative Paths

The following two exercises should each take about 30 minutes. They can be tackled by individuals, or in groups of three students, to promote discussion and obtain feedback about the principles involved.

Question 1

Two ships are sailing on steady courses in the open sea. At 08:00 an observer on ship A, which is travelling due North at 25 km/h, sights ship B on a bearing of 60°, a distance 8km away. Ship B is moving at 50 km/h due West.

a. Show their actual positions and velocities on a space diagram, drawn roughly to scale.
b. Determine the velocity of B relative to A.
c. Sketch the relative path of B as seen from A, approximately to scale, indicating all known dimensions and angles.
d. Determine the closest distance between the two ships if they continue without changing course.
e. Determine the time at which they will reach the point of closest approach.
f. What will be the relative bearing of ship B as seen from ship A, at 08:06?

[V_{BA} = 55.90 km/h W 26.57° S; closest distance = 479.4 m, reached at 08:08:34; relative bearing is 52.11°]

Question 2

Ship A is moving on a steady course, on a bearing of 42°, doing 30 km/h. At 12:00 the lookout detects another ship, B, on a bearing of 15°, a distance 6 km away. Seen from ship A, ship B appears to be moving due South at 25 km/h.

a. Draw a position diagram showing clearly the positions of both ships at 12:00.
b. Determine the actual velocity of ship B, in magnitude and direction.
c. At what time will ship B come within 2km of ship A?
d. For how long will ship B remain within 2 km of ship A?

[22.83 km/h; S 77.54° E, or E 12.45° S; 12:10:53; for 6 minutes and 3 seconds]

Intercept situations

In all the relative path examples we have seen, we have been able to determine the distance between the two moving objects at the point of closest approach between them. To re-cap, we show below the relative path of ship A as seen from ship B:

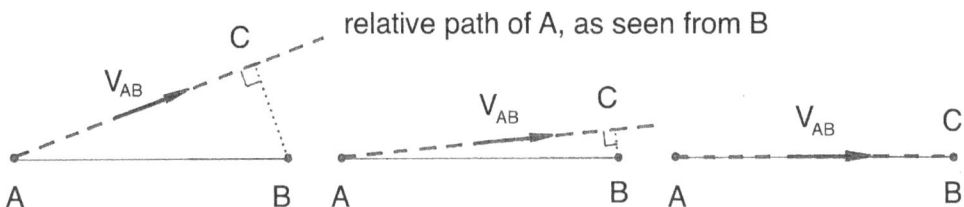

relative path of A, as seen from B

The shortest distance between the two ships, at the point of closest approach, was the distance from B to the relative path. (Distance BC.)

Now, if that shortest distance becomes zero, interception will occur. This will occur if the relative path passes through B.

Therefore, for an object A to intercept an object B, the vector v_{AB} should point in the same direction as the line AB.

Example

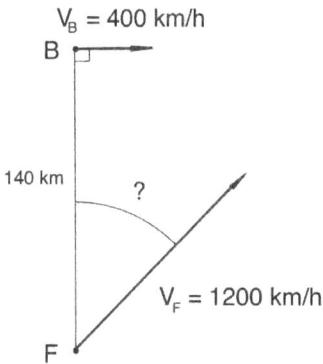

A fighter plane, with a cruising speed of 1200 km/h, takes off to intercept a bomber which is flying due east at 400 km/h.

At the moment of take-off, the bomber is at a point exactly 140 km north of where the fighter is. In what direction should the fighter plane fly, in order to intercept the bomber?

For intercept to occur, V_{FB} must be directed along line FB. So, V_{FB} must run due north.

Write the vector equation for V_{FB} :

$$V_{FB} = V_F + (-V_B)$$

Now, sketch the vector diagram that the equation describes:

From the diagram, sin α = 400/1200

therefore α = 19.47°.

Hence, the fighter must fly on a bearing of 19.47°

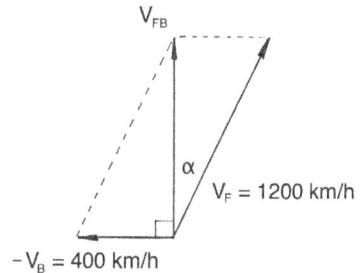

Exercises with Intercept Situations

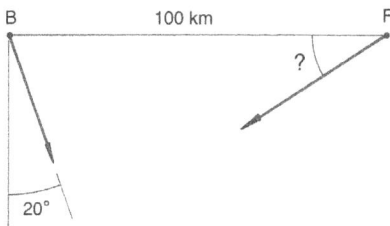

Question 1

A bomber, B, leaves its base, flying at 1200 km/h in a direction S 20° E. At the same instant, an enemy fighter plane, F, takes off from its own base, to intercept the bomber. The base from which the fighter leaves is 100 km due east of the bomber's base. The fighter is capable of 1600 km/h.

a. In what direction must the fighter fly in order to intercept the bomber in the shortest possible time?   [W 44.81° S]

b. After what length of time does F intercept B?   [3 min 53 sec]

Question 2

A Viking ship is being rowed downstream in the centre of a river that is 200 m wide. The ship is moving at 5m/s relative to the water, which is flowing at 2m/s. An enemy archer is waiting at a point A on the bank of the river. He decides to shoot at a particular member of the ship's crew. The velocity of his arrow is 80 m/s.

a. If he waits until the ship is directly opposite him before shooting, how far ahead of the target should he be aiming when he releases the arrow?   [8.784 m, hence approx. 9 m]]

b. How far does the arrow actually travel?   [100.39 m]

c. How long will it take the arrow to reach the target?   [1.255 sec]

Question 3

30 km/h   50 km/h

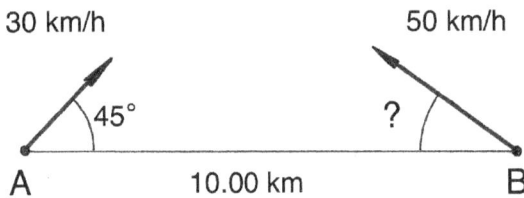

45°   ?

A   10.00 km   B

At 12:00, a suspicious-looking ship, A, is detected by a coast guard vessel, B.

The ship is 10 km due West of the coast guard vessel, and is cruising at 30 km/h on a bearing of 45°. The coast guard vessel is capable of moving at 50 km/h.

a. In what direction should ship B head, in order to intercept ship A in the shortest possible time? State this direction as a compass bearing. [295.1°]

b. At what time will it intercept?   [12:09:02]

c. What is the actual distance covered by the coast guard vessel from the time of sighting the ship to the time of interception? [7.520 km]

Question 4

10 m/s

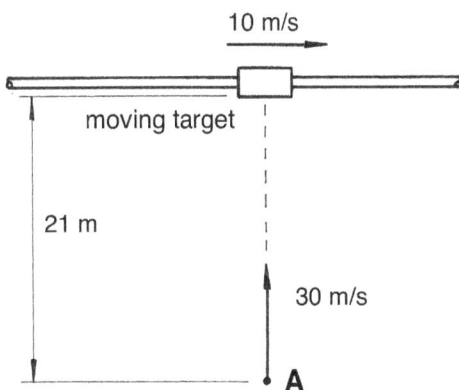

moving target

21 m

30 m/s

A

The diagram shows a plan view of a venue used for archery practice. An archer in position A shoots an arrow intended to hit a moving target that is propelled mechanically, with constant velocity of 10 m/s, along a straight horizontal rail running from his left to his right. He is standing 21 metres from the rail. The velocity of the arrow is 30m/s.

a. If he wishes the arrow to hit the target at the instant that the target is closest to him, how far to the left of that position should the target be when he releases the arrow?   [7.000 m]

b. If he fires at the instant that the target is closest to him, how far to the right of the target should he aim? [7.425 m]

c. In case b. above, how far to the right of the closest position will the target be when the arrow strikes it? [*deliberately left unanswered, for discussion*]

Relative velocity methods applied to basic mechanisms

The purpose of a mechanism is to convert an input consisting of one kind of motion, to an output with a different kind of motion. This is achieved by using a sequence of links that either pivot or slide or roll in relation to other links.

For example, the well-known mechanism consisting of a crank, connecting-rod and piston in a cylinder, allows rotary motion to be converted to reciprocating linear motion, or vice versa.

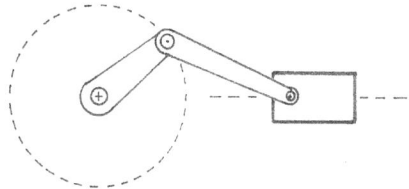

Designers of machinery need to be able to determine the velocity of an output motion, given the velocity of an input motion. They also need to be able to determine the accelerations, and hence the forces, on all the links in a sequence. For the purposes of the present chapter, we will not analyse accelerations, but will confine our analysis to velocities.

It will become evident that the relative velocity equation that was developed earlier in this chapter, and the vector diagram that this equation describes, are essential in determining unknown velocities in a mechanism. For example, consider the following mechanism:

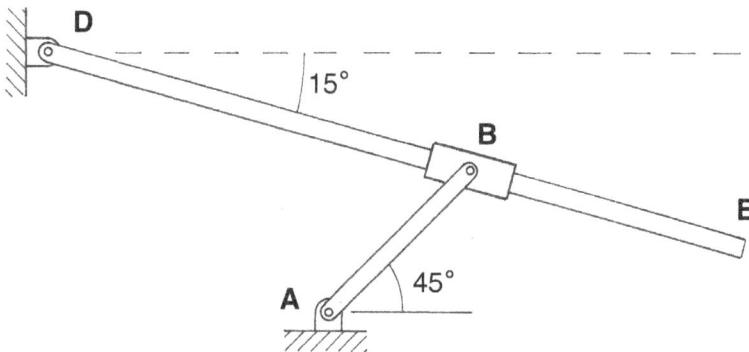

Points A and D are fixed in relation to the frame of the machine. Link AB, which is 200 mm long, is driven to rotate clockwise with a uniform speed of 12 rad/s. The end B of this link is pivoted to a collar that slides on rod DE. Assume that the movement of link AB does not impede the range of movement of rod DE.

It is easy to imagine the type of movement of rod DE: an up-and-down sweeping motion in a portion of an arc, with variable angular velocity.

Suppose we need to determine the velocity of point E at an instant during the motion when the mechanism is in the position shown. In this position, DB = 450 mm and DE = 750 mm.

We reason as follows:

To determine the velocity of point E, we need first to determine the velocity of that point on the rod which at the given instant is inside the collar, level with point B. Let this point be called R (for rod).

So, we need to determine V_R, whose value is not immediately apparent.

The only velocity that is defined, to begin with, is that of end B of link AB. This velocity is

V_B = rω = 0.2 × 12 = 2.4 m/s, perpendicular to AB, as shown here.

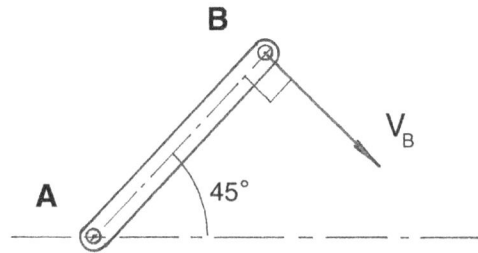

Now, points B (on the driving link) and C (which we will name the pivot on the collar) are physically constrained to move together. Thus $V_C = V_B$.

The method of determining V_R is to relate three vectors which form a relative velocity vector diagram. Since we know V_C and we wish to find V_R, we will use the vector equation: $V_{RC} = V_R + (-V_C)$, where V_{RC} is the velocity of the rod relative to the collar.

Instantaneous directions of

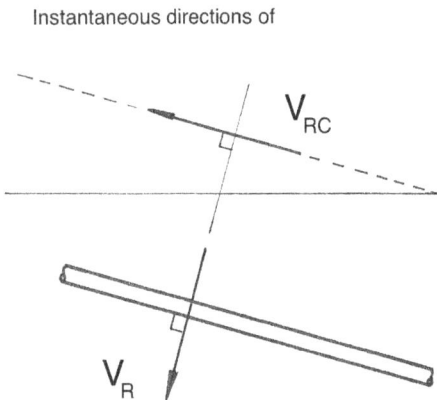

Although we don't yet have the magnitudes of either of these two unknown velocities, we can obtain their *directions* by inspecting the diagram of the mechanism in the given position.

V_{RC}, the velocity of the rod relative to the collar, has to lie in the same direction as the centreline of the rod at the instant being investigated. By inspection, we can see that the collar will be moving to the right on the rod, so V_{RC} will be directed to the left, and will therefore lie in the direction W 15° N.

V_R has to be perpendicular to the centreline of the rod, as the rod is pivoted to the machine frame, and thus point R is constrained to move in a circular arc with pivot D at its centre. Furthermore, by inspection, we can see that point R must be moving downwards at the instant under consideration. This defines the direction of V_R.

Having one vector completely specified, and knowing the directions of the two others, we can arrange the vector diagram as follows:

Since V_R and V_{RC} are at right angles to one another, the vector diagram is a right-angled triangle, and is easily solved.

From the diagram, the magnitudes of the two previously unknown vectors are:

$V_{RC} = 2.4 \cos 30° = 2.078$ m/s

and

$V_R = 2.4 \sin 30° = 1.2$ m/s

By proportion, $V_E/V_R = 750$ mm $/450$ mm

$\therefore V_E = 2.0$ m/s

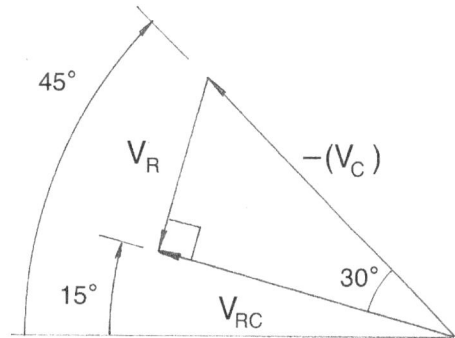

Exercises on relative velocities in mechanisms

Question 1

Sand coming off conveyor belt A hits belt B with a velocity of 3 m/s at 30° to the vertical. It is desired that the relative velocity of the sand to belt B should be at right angles to belt B (otherwise the stream of sand will abrade the belt).

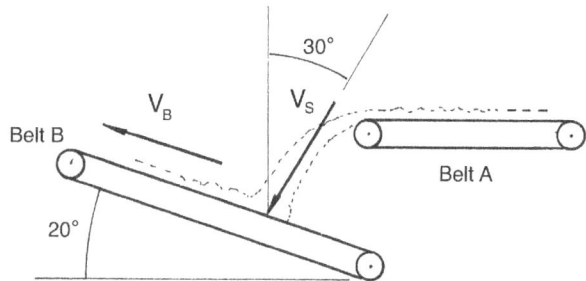

Determine the required velocity of belt B to achieve this.

[0.52 m/s]

Question 2

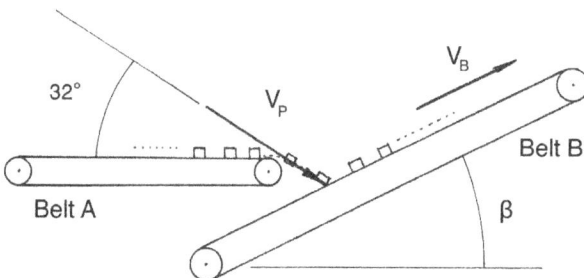

Packages from conveyor belt A strike belt B with a velocity of 2.1 m/s at 32° to the horizontal. Belt B moves at 1.2 m/s.

Determine angle β so that there will be no slip between the packages and the belt when they land.   [23.15°]

215

Question 3

A wheel of diameter 1000 mm is rolling from left to right, without slipping, on a straight level floor, at 2 m/s. Consider the direction to the right to be positive. Consider three points on the wheel: one at centre A, one at point B and the other at a point C on the rim, directly above the centre of the wheel.

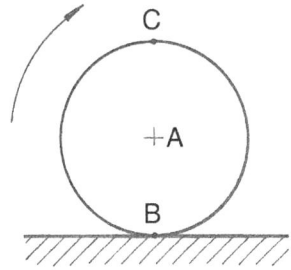

At the instant when point B is in contact with the floor, state the value and direction of each of the following (*answers not supplied: the exercise is suitable for discussion in a group*):

The velocity of A relative to the floor (V_{AF});

The velocity of B relative to the floor (V_{BF});

V_{AB}; V_{BA}; V_{CB}; and V_{CA}

Question 4

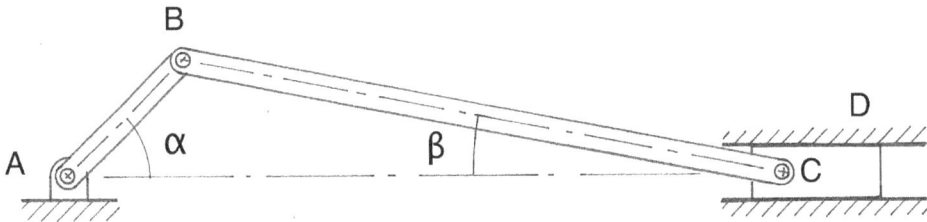

Shown here is the essential arrangement of a crank AB (160 mm between centres), joined by a connecting rod BC (600 mm between centres) to piston C. Crank AB is driven to rotate clockwise at a constant angular velocity of 24 rad/s. Determine:

- The instantaneous velocity of the piston within cylinder D when angle $\alpha = 40°$. [3.371 m/s to the right]
- The minimum instantaneous velocity of the piston at any point in its cycle of movement, and the value of α when that occurs. [0 m/s; at $\alpha = 0°$ and 180°]
- The instantaneous velocity of the piston when AB is perpendicular to BC. [3.974 m/s]

Question 5

Consider the mechanism shown below: Rod AB pivots on the machine frame at point A.

Rod FC is pivoted to this rod at point C, and is driven by the wheel with centre G, which rotates anticlockwise at a constant speed of 10 rad/s.

Collar D is pivoted to rigid link DE, which is pivoted to the machine frame at E.

AC = 200 mm;   FC = 250 mm;   DE = 300 mm;   FG = 25 mm

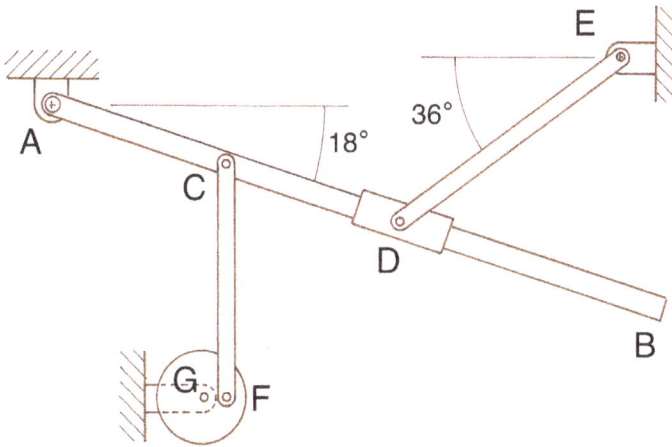

At the instant that the mechanism is in the position shown, rod FC is vertical and AD = 400 mm.

Determine the instantaneous angular velocity of link DE in this position.

[2.891 rad/s clockwise]

General questions on relative velocity

	Identify whether each of the following statements is true or false		
1	All velocities are relative.	T	F
2	What we call a 'true' velocity is actually a velocity relative to the Earth.	T	F
3	The relative velocity of A to B is the velocity that A appears to have, as seen from B.	T	F
4	If an object, P, is moving in a medium, Q, that is also moving, then, to determine the velocity of P relative to Earth, we must subtract the velocity of Q from that of P.	T	F
5	A resultant velocity is a velocity relative to the Earth.	T	F
6	An aircraft needs to fly from one airfield to another that is due east of it. The cruising speed of the aircraft is 240 km/h. There is a wind coming directly from the north, blowing at 80 km/h. Under these conditions the ground-speed of the aircraft will be greater than 240 km/h.	T	F
7	The relative path between two moving objects is a real path that can be traced on the ground.	T	F
8	The velocity of A relative to B may be found by determining the resultant of the following two velocity vectors: V_B and $(-V_A)$	T	F
9	For two objects, A and B, moving in the same plane, the velocity of A relative to B is always equal and opposite to the velocity of B relative to A.	T	F

Index

Acceleration...3
Acceleration Ratio 58, 61
Air Resistance 21, 22, 40
Air-bed rail 160
Angles of friction and repose 128
Angles of incidence and departure.......189
Angular Acceleration................................51
Angular and linear counterparts 50, 53
Angular Displacement 48
Angular Velocity.......................................48
Arc Length 46
Archimedes' screw 95
Area under velocity-time graph.................8
Arrow, motion of..................27, 29, 30, 172
Ballista.......................................34, 43, 44
Ballistic Pendulum ....165 - 167
Billiard ball collisions....................157, 181
Block and tackle systems .... 113 - 117, 131
Bullets...................................162, 165, 169
Cannon Ball....................................165, 166
Catapult 41
Chain Hoist....................................126, 135
Chains and Sprockets 60
Coefficient of Restitution..............176 - 179
Collisions in one dimension..................157
Collisions in two dimensions........179 - 183
Collisions, elastic and plastic...............174
Collisions, soft 159
Collisions, vehicles 180, 183
Compass bearing 197
Complex Machines..................................104
Compound gears.....................................59
Connecting rod 216
Conservation of Energy..................82 - 87
Conservation of Momentum 153
Constant speed, calculations for 5
Conveyor belt, angle of incidence 186
Crank and differential axle....................133
Deceleration 1, 4
Design-and-Build Projects
...............34, 38, 39, 42, 43, 134, 136, 167
Dimensions...1
Displacement...2
Displacement-time Graph........................16
Distance 2

Drag Coefficient.......................................22
Earth's Radius 63
Efficiency of a Lifting Machine..... 119 - 122
Effort, definition.....................................105
Elastic Energy..............................76, 79, 80
Energy Accounting Diagram...........85, 108
Energy lost in an impact 159, 175
Energy stored in a coil spring 75
Energy stored in an archery bow...........74
Equations of Linear Motion............. 11 - 13
Flight path of projected object 25
Fluid Stream, momentum of ....184 - 190
Fluid Stream, force exerted by 185
Friction wheel 58, 65
g Forces..57
Galileo 21
Gears..58
Gearwheels, Meshing........................... 111
Gradient of velocity-time graph.................7
Gravitational Acceleration.................21, 22
Gravitational Potential Energy...............77
Gravity 21
Grindstone, Hand-cranked 96
Hooke, Robert 69
Hooke's law 69
Horsepower, definition............................88
Hose Nozzle 184, 185
Human cannonball example..................81
Impulse and impulsive forces ......169 - 171
Inclined Plane, as machine component
.... 104, 111
Inertia of linked masses.......139, 142 - 146
Inertia, Nature of...................................140
Intercept situations, velocity for 210
Kinetic Energy 78
Launch velocity, determining 21, 38
Law of a simple lifting machine............ 118
Lever...104
Linear Accelerating Systems 139
Linear Motion..1
Load height raised........................106, 107
Load, definition 105
Loading, gradual, sudden and shock
....97 - 101
Mangonel..37

Mass flowrate 185
Mechanical Advantage 105
Mechanical forms of energy 76
Mechanical Work, Definition 67
Momentum Transfer 153
Momentum, Law of conservation of..... 153,
....155 - 157
Momentum, Linear 151, 152
Motion, implications for engineers 1
Motorcycle 17, 41
Negative direction.................................... 4
Newton's Cradle 158
Newton's Third Law 154
Newton's Second Law 140
Nut on screw thread, lifting device........ 112
Obelisk, raising of 103
Piston.. 216
Pitch Circle 58, 59
Point of Closest Approach 205
Positive direction 4
Power output, instantaneous.................. 91
Power transmitted by a torque................ 93
Power, average and maximum 90, 91
Power, calculation examples 89, 90
Power, definition 87
Power, amount exerted by human beings
.... 88
Projectile motion.............................30 - 37
Projectiles, equation of the trajectory 36
Projectile, horizontal range 35
Pulley, idler or tensioner 61, 64
Pulleys and belts 60
Radian as a measure of angle........45 - 48
Radius of Earth...................................... 48
Rangefinder, on rifle 38
Recoil... 162
Reference level for potential energy 23
Relative bearing, in directions 197, 198
Relative path of a moving object .204 - 210
Relative velocity......................see Velocity
Relative velocity in mechanisms..213 - 217
Revs per min conversion to rad/s 49
Rifle adjustment.............................38, 39
Robins, Benjamin 165
Rollers 104
Rolling resistance 84, 85
Rope tension, variation in 127
Rope, effect of mass............................ 146
Rotational Inertia 139

Rotational Motion............................45 - 66
Rubber Band, force vs extension 75
Scalar 3
Screw operated jack 124
Sheave, compound or differential......... 111
Sheave, definition 104
Sheave, standing and running...... 109, 110
Simple lifting machine, safety
considerations 122
Simple Lifting Machines, definition 104
Speed 3, 5
Sports balls.................30, 32, 40, 153, 174
Spring, force vs extension 69
Steam engines...................................... 88
Stiffness coefficient of a spring............... 75
System of objects 155
Terminal Velocity.................................... 22
Thrust, of a jet or rocket engine............ 184
Time delay, launching of two projectiles
....27 - 29
Trajectory............................21, 34, 35, 36
Trajectory, of golf ball........................39, 40
Trampolinist, movement of 42
Treadle-powered grindstone.................. 53
Useful work output of a machine 109
Vector 3
Vectors, describing directions of........... 196
Velocity diagram for relative velocity
....201 - 204
Velocity Ratio, of a lifting machine
....106 - 109
Velocity Ratio, of gears.......................... 58
Velocity-time Graph ....7 - 10, 54 - 56
Velocity, absolute................................. 193
Velocity, definition 3
Velocity, relative...................194, 199 - 204
Velocity, resultant................................. 195
Vertical Motion 21, 23
Water craft, human powered 189
Waterwheel.. 80
Watt, definition of 88
Watt James.. 88
Wedge 104, 128
Wheel and differential axle 125
Wheel, historical record of..................... 45
Winch, Geared............................. 134, 136
Winch, Hand-cranked..............51, 94, 134
Work Done against a force 71
Work Done, as area under graph 68, 73

Work done by a force 67
Work Done by a torque........................... 92
Work Done by a varying force 68, 74
Work Done by an oblique force 70

X-Y co-ordinate system 30
Zero Energy Losses, hypothetical 108
Zeroth Law of Mechanics 71, 154

About the Author

Gregory Pastoll PhD (Higher education, University of Cape Town, 1994) BSc Mechanical Engineering (University of the Witwatersrand, 1973) has had a 29-year career as an academic, shared equally between two fields, mechanical engineering instruction and teaching methods in higher education.

He has been a lecturer in mechanical engineering at The Cape Technikon, Cape Town, also at The Cape Peninsula University of Technology, Bellville, and senior lecturer in teaching methods at The University of Cape Town.

He taught basic engineering mechanics in a polytechnic/further education environment for over 14 years, being the subject co-ordinator of Mechanics 1 for some years and for Mechanics 2 for some years. He re-introduced lab experimentation into the teaching of mechanics in his department and started a tradition of using design-and-build projects for students to get hands-on experience of the principles they were supposed to be learning in class.

The author has also obtained teaching experience in two language academies, teaching English as a foreign language in South Korea (1 year) and Austria (2 years).

The author adjudicating student projects. In this case students had to design and build vehicles that had to travel as far as possible on a level floor, using a specified amount of energy supplied by a mass-piece descending through a height.

Photo by an unknown student, one of many assisting at projects such as these.

His other educational books are:

'Motivating People to Learn…and teachers to teach' (Authorhouse, 2009)

'Tutorials That Work' (Arrow Publishers, 1992).

The author has written and published fictional works, including children's stories, short stories in rhyme for adults, and the scripts for five musicals, two of which have been produced by a primary school. He has built three violins and designed and marketed a board game based on the sport of cricket. His hobbies are painting and woodwork.